# The Wonders of Salvage

*by*

David Masters

WITH FORTY-EIGHT ILLUSTRATIONS FROM PHOTOGRAPHS

*First Published in 1924*

TO
MY WIFE

NATAL PUBLISHING LLC
*ARS LONGA, VITA BREVIS*

Hopeless as the S.S. *Devona*'s position seemed on September 15, 1917, the salvors managed to raise her in four days. Very cleverly they rigged up some wire mattresses into which they pumped her cargo of wheat, thus draining off the water and saving the grain.

# LIST OF ILLUSTRATIONS


Wreck of S.S. *Devona*

Examining Sea-bed in Tobermory Bay

Washing Sand for Signs of Treasure

Sifting Sea-bed for Gold of *Lutine*

Wreck of *Oceana*

Diving for *Oceana's* Treasure

A Diver Treasure-Hunting with Explosives

Bringing the *Leonardo da Vinci* upside down into Dock

The *Leonardo da Vinci* safely docked

The Mammoth Timber Framework on which the Upside-down Battleship rested

The Upside-down Battleship seen from the Air

Towing out the Upside-down Battleship to turn her over

The Battleship just before she was righted

The *Leonardo da Vinci* as she swung over

The Battleship righted

A Torpedoed Ship safely beached

The Famous Standard Patch

Electric Pumps in the Hold of a £3,000,000 Ship

Damage wrought by a Torpedo

A Vessel down by the Head

The *U-44* carried Ashore

Removing Mines from the *U-44*

The *K.13* raised after Two-and-a-half Days on the Sea-bed

A Blazing Oil Tanker

The *Onward* overturned at Folkestone

Salvage Craft alongside the *Onward*

Tug-of-war between Five Railway Locomotives and an Overturned Troopship

Pumping Out the *Onward*

Wreck of *St. Paul* in New York Harbour

Overturned Liner beside the Quay

Dragging the *St. Paul* upright

The *St. Paul* raised

The *Araby* blocking the Entrance of Boulogne Harbour

The *Araby* breaking in Two

Two Halves of the *Araby* beached

Half a Ship in Mid-Channel

Patching a Ship with Concrete

How the Concrete Patch was reinforced

Concrete Patch from Inside the Ship

Exterior View of Ship patched with Concrete

Refloating a Wreck by Digging Operations

The *Timbo* High and Dry

A Dredger wrecked in the Gareloch

Mighty Steel Cables used for righting the Wreck

The Dredger righted Once More

A Torpedoed Ship in Grave Difficulties

The Foundering Ship safely beached at Clovelly

Salving a Wreck from Quicksands

# CHAPTER I

With eyes gazing fixedly ahead, the man, tense and alert, sought to penetrate the blackness. Squalls of rain swept down and lashed his face, the flying spume of spray shot up to intermingle with the rain, leaving a tang of salt on his lips. The liner lurched and rolled through the night, while thousands of souls aboard slumbered without fear, placing implicit trust in this one man to whom the pulse of the engines driving the ship was as familiar as the pulse of his own heart. Rain and spray and wind were part of his life, and he accepted them without demur because he realized that the weather was indifferent alike to praise and blame.

He half turned his head to glance at the ship's chronometer.

"Should be picking her up now," he muttered.

Raising the night-glasses to his eyes, he concentrated all his powers of vision on the murky gloom in front of him. His glasses roved slowly from side to side, then a point of light, so dim as to be almost imperceptible, swung in the blackness and vanished. For a minute he waited until the light reappeared, then he breathed freely and rang down for the ship to alter course, knowing that he was safe and that he had justified the faith of the passengers who had trusted him to navigate his vessel through the storm.

That point of light which meant so much to him was the beam of a lighthouse, one of the many encircling our coast. All round our shores they keep sentinel night after night, through summer calm and winter blizzard, waking to life as daylight fades and dying as dawn steals over the seas. These lights, which the city dweller on a brief visit to the sea watches with such interest, are the friends of all who go down to the sea in ships.

Our coasts are profoundly treacherous. Rocks, shoals

and quicksands abound everywhere, and are mostly marked with lighthouses, lightships and buoys which in the aggregate have cost millions of pounds. No expense has been spared to indicate these hidden dangers and make our seas safe for shipping. Yet, in spite of all that human foresight can suggest, wrecks still occur. Gales spring up and take their toll; fogs steal on and drive ships blindly to their doom; machinery breaks down and allows the seas to hurl the helpless craft upon the cruel rocks.

Probably no coast in the world is so well lighted as that of Great Britain, but although there are over 1700 lights acting as signposts of the sea, warning mariners of their dangers, our rocky shores exact a grievous toll of shipping year by year. It is estimated that the average value of the ships and cargoes lost in British waters amounts to about £5,000,000 annually, so the wealth spilled out of the ships since the galleys of our first invaders found a watery grave would, could it be recovered, considerably lighten the burden of our national debt. Unfortunately the greater part is lost for ever, for the sea which has swallowed the ships destroys them utterly in the course of time, and unless they can be salved within a certain period they soon become not worth salving. The action of the sea water rots away the cargoes, rust gradually devours the steel and iron carcass of the ship, and only those two indestructible substances, gold and silver, the white and red metals for which men have fought and died throughout the ages, remain of the wealth which was originally lost.

Men, however, have not been content to see fortunes sink in the sea without making some effort to recover them. They have pitted their wits against the strength of the sea, risked their lives to wrest long-lost treasure from the grasp of the ocean, and the story of their thrilling deeds is one of the outstanding pages of human endeavour.

Consider, for a moment, the wonder of a ship. She is a marvellous structure of steel and iron, full of the most intricate machinery, a structure weighing perhaps thousands of tons. Of the manifold parts of which she is composed, the wood fittings alone may be buoyant. Only they may possess the power of floating on the waves; all the other parts, from the smallest screw and rivet to the mighty propeller shafts and hull plates would, if they could, sink like stones to the bottom of the sea. This enormous mass of metal, which in its natural state must sink, is so cunningly fashioned by man that it overcomes its natural inclination to sink and is made to float. The huge weight is supported by water, men toil in the bottom of the ship 20 and 30 feet below the surface of the sea and are oblivious of any danger. The water on the outer side of the steel skin of the ship towers 20 and 30 feet above their heads, yet they sleep and eat and work in perfect safety. So long as the sea is prevented from washing over the sides of the ship or entering through a breach in the hull the vessel floats, would continue to float even were she made of lead. In other words, she is buoyant. Only when her buoyancy is destroyed does she sink. Then, before she can float again, her buoyancy must be restored.

This is the simple problem that is always confronting the sea salvage expert. How can he restore the buoyancy of the ship that meets with misfortune? Simple as is the problem, it is seldom that the answer is easy. To the salvor every wreck is a riddle. Tides and currents make the riddle more complex. The position in which the wreck is lying profoundly affects the case. And, above all, operates the unknown factor of the weather. Whatever the salvage expert hopes to do, he always adds to himself "Weather permitting!" He may be the cleverest man alive, his plans of salvage may be the most brilliant ever conceived, he may have the most expensive plant at his disposal and all the money he seeks at his command, yet he is helpless unless the weather be fair. Plans may be put into operation, work may go smoothly, everything may be within an ace of success—when the tail of a gale may blow the plans to pieces, shatter the work and rob the salvor of the success that seemed within his grasp. It has happened before many times, and it will happen many times again.

The men who get a living by trying to raise wrecks are farseeing, sparing of words, patient where patience is demanded, quick as a rapier thrust where quickness is essential, capable of toiling until they drop if it be necessary. Every contingency that it is possible to think of they consider, but the weather is something beyond their control. They pray for fine weather, and fight against foul to the best of their ability; but when the wind takes hold man and his endeavours are as nothing.

Hard as some of the salvors have worked for their successes, others have worked harder still for their failures. Often and often they have striven strenuously for weeks and months to salve a ship, only to lose her in the end. The luck of the game is indicated by a case which occurred a year or two ago. A vessel went down on the summit of a rock jutting sheer from the seabed. On all sides was water so deep that she had but to slip to be irretrievably lost. The salvors, hurrying to the scene, found her balanced most precariously on a ledge. A glance told them that, before they could make the slightest attempt to salve the ship, they would have to strive their utmost to secure her firmly in position on top of the pinnacle of rock. They routed among their gear for cables and anchors and, making the cables fast to the ship, carried out the anchors in all directions in order to tie her tightly into place.

Then they began to work against time, keeping a keen eye on the sky and praying for fine weather, knowing full well that if the weather held fair they would save the ship and that the coming of bad weather would seal her doom. Day after day they toiled like giants, struggling with huge baulks of timber, shoring up decks, strengthening bulkheads, patching breaches in the hull. The weather favoured them. Day after day it remained fine and enabled them to carry on their operations quite unhampered. They had been hard at it for nearly a month before the breeze began to freshen in rather an ominous manner. They were just beginning to anticipate rough weather when the wind luckily died away and they breathed freely once more.

They redoubled their efforts, and six weeks of intense toil saw their work completed. The last timber was bolted securely in place and the divers came out of the wreck, announcing that all was ready for pumping out on the morrow. The salvors turned in for the night well pleased with their labours, conscious that the next day would see them proceeding to port with their prize.

But the weather, which had been kind to them so long,

was destined to cheat them at the very last. That night it began to blow. The seas started to rise and hammer at the ship. She began to stir uneasily and to strain at her cables. The gale increased. Under the continuous chafing, one cable suddenly snapped. The breaking of that cable gave the wreck more freedom to move under the hammer blows of the sea. The waves battered at her incessantly and one cable after another went like threads of cotton until a billow, far mightier than the rest, caught her up and swept her off the pinnacle into the depths.

Imagine the feelings of the salvors when day dawned. All their gear was gone, their labours lost when the prize was within their grasp. They steamed slowly round the spot and proceeded to port, hoping for better luck next time. That was the only thing they could do.

Men who spend their lives on salvage work are rather apt to lead the casual inquirer to imagine that it is the easiest job under the sun, whereas in reality the task is beset with difficulties and bristles with risks. But the sailormen in their matter-of-fact way forget to mention the ever-present danger. They are inured to it, just as people are habituated to living on the slopes of a volcano that may erupt and overwhelm them at any moment of the night or day. None the less the salvors never forget the risk, nor leave it out of their calculations, and for this reason fatal accidents among them are rare. They know the strength of the sea too well to attempt to take liberties with it, for they have seen it pick up great 10,000 ton ships and toss them on the rocks as though they were cockle-shells; they have seen the strength of 70,000 horses in the engines of a ship struggling in vain against the strength of the waves, and they know better than to pit their power against the power of the storm.

Thus they have a wholesome respect for wind and wave. They use the strength of the sea to further their own ends so long as the sea permits. At other times they may stand by a wreck for weeks while the sea seethes and the wind howls about the ship they seek to save. A lull in the bad weather will set them working frantically, and more than one ship now afloat owes her existence to the accumulated labour of a number of short spells of work undertaken between the gales.

The salvage man must thus be infinitely patient and possess a determination that will keep him at work when most other men would give up in despair. Above all must he be strong in hope. Without hope, no man need seek to become a salvage expert, for he would be foredoomed to failure. He must possess not only physical courage that enables him to face the dangers of his calling, but also that rarer mental courage that enables him to snatch victory out of the very jaws of defeat.

It is the men who possess this mental as well as physical courage who perform the wonderful feats of salvage that will never be forgotten, such as the recovery off Gibraltar of the steamer *Hypatia*, which the salvors brought to the surface after an infinity of trouble. No sooner was she raised than she filled and sank like a stone.

There was nothing for it but to do the work over again, which the salvors managed to do. For the second time the *Hypatia* was brought to the surface, and once again she sank, seeming to mock the efforts of her would-be preservers. Still they were not beaten. With grim determination they made another effort, and after a great fight managed to raise the *Hypatia* once more. All in vain! For the third time she sank.

Notwithstanding these three reverses, the salvors would not give up the fight. Again the divers went down, and their strenuous exertions ended in the *Hypatia* seeing the light of day yet again. Not for long were the salvors allowed to rest after their labours. Down she went for the fourth time, while the sea bubbled and boiled around.

Few men would have continued a fight which appeared so hopeless. But the salvors would not admit themselves beaten. Although Fate seemed to be taunting them, they had the courage to take their task in hand for the fifth time, and this time they succeeded. Truly it can be said that no men more fully earned their reward than these salvors who triumphed after four defeats.

# CHAPTER II

From earliest years our imaginations are fired by the mere mention of treasure. Who has not heard of that fabulous treasure of the bloodthirsty pirate, Captain Kidd, whose booty still lies hidden on some far-off island? Expedition after expedition has been fitted out to find it, but the pirate hid it so well that the hunters have failed in their quest. Who has not marvelled at those mighty hoards of gold stored away by the Incas of Peru, gold which Pizarro looted from the Peruvian treasure-house and carried back to Spain?

Treasure! The mere whisper works magic, conjuring up pictures of gold and silver and piles of glowing gems—rubies, emeralds, and diamonds galore, gleaming with all the colours of the rainbow. So fascinating is the idea of treasure that men gladly risk their lives to go in search of it; nor is the magic confined alone to the romantic. The keenest of business men, who boast of their hard-headedness, seem to lose their heads where treasure is concerned. Eagerly they fling down the funds to prosecute the most problematic searches, in return for the promise of the most shadowy spoils.

These same business men will aver that they never speculate, yet all treasure-hunting is speculative, and if there is one form more speculative than another it is that of searching for sunken treasure. Still, despite its hazardous nature, there is always money forthcoming to back deep-sea enterprises of this description. True, success comes but seldom—failures are the rule. Could a correct balance-sheet be made up showing how much has been spent on hunting for the world's sunken treasure and how much has been recovered, we should probably find that the money expended was many times greater than the value of all the treasure brought to the surface.

Few ideas could be more fascinating than that of hauling

up gold and silver from the bottom of the sea, and it is this same fascination, with all the excitement it brings in its train, which lures men on to attempt to wrest many of these long-lost treasures from the recesses of the ocean. Years sometimes are spent in pondering ancient documents, hunting for evidence of the exact locality of the vanished treasure, seeking to sift rumour from actual fact. Further years may be spent in making plans and special apparatus for lifting the treasure, and, when the hunter starts in real earnest, he finds at last that he has spent years of his life and thousands of pounds just for the privilege of stirring up the seabed. Treasure-hunting is, in fact, something like taking a ticket for a sweepstake. The chances may be ridiculously small, but the prospect of winning a fortune will always make the game popular.

Fate, indeed, seems to delight in playing tricks on salvage men. While, on the one hand, it sometimes leads them on to fit out ambitious expeditions costing thousands of pounds, sends them journeying afar and imposes the greatest hardships upon them without bringing them any reward whatsoever; on the other hand, it sometimes flings a fortune straight into the lap of some lucky man when he is least expecting it.

Lord Leverhulme, in illustrating the vagaries of Fate, related how an Australian firm once owned an island in the Pacific, a rocky little place with a few coco-nut trees that gave their crop of nuts which were duly dried in the sun and turned into copra and coco-nut oil. Their trading schooner used to visit the island to load the copra, and on one of the trips the captain happened to pick up a piece of rock and put it aboard the ship. In due course that piece of rock went back to Australia with the copra, and was used in the office to keep the door open when the weather was sultry.

The firm acquired their island to make money out of it, but although the coco-nut trees brought them a profit, they certainly did not bring them a fortune. The question arose as to whether it was worth their while retaining the island, and after due consideration they sold their property to some one else, and thought no more about it.

Entering their office one day, a professor from the

university chanced to kick against the stone that was propping the door open. He stooped down, picked it up, scrutinized it closely for a minute or two.

"Where did you get this?" he demanded.

"Oh, that's a bit of rock our skipper brought back from one of our islands," was the reply.

The professor looked at the rock again. "Do you know what it is?" he asked.

"Just a bit of stone," came the answer.

"I don't know," said the professor, "but I think it's phosphate. I'd like to take it away and analyse it, if you'll allow me."

Permission was, of course, granted, and the professor walked away with that bit of rock which scores of men had kicked against at the door. Taking it to his laboratory, the scientist carefully analysed it. He found it to be a sample of the richest phosphate in the world. The original owners had bought the island as a business proposition, but they failed to realize the fortune that was theirs. That rocky island turned out to be one mass of phosphate, worth about £100,000,000—and they had let it go for a few hundreds! Of all who had stumbled over that lucky door-prop, the professor was the only one who had the sense to see the fortune lying at his feet.

The counterpart of the professor who saw a fortune in that neglected lump of rock was the diver who heard the whisper of truth in a rumour. The work of this diver took him to the coast of Galway, where he was engaged on salvage work that was to last some little time. He was a companionable sort of man and, after finishing his spells of work, would adjourn to the tap-room of the village inn to spend his evenings in yarning with the fisherfolk.

For years a story had been current in the neighbourhood that a Spanish galleon, one of the ships of the Armada, had gone down in the vicinity. Those who heard the yarn smiled. "It's just a rumour," they remarked.

Whether it was merely a rumour, or something more, the story had been told from father to son for generations. So persistent a rumour was it that it survived century after century, living in the traditions of these simple Irish fisherfolk, passed on by word of mouth in the little community, until it survived to our own times. Most of the fishermen knew the yarn of the sunken Spanish galleon, but perhaps the passage of time had made many of them rather sceptical.

Anyway, one evening the diver was enjoying his pipe and his beer and talking about his work, when an old fisherman said to him:

"Why don't ye thry for the galleon?"

"What galleon?" the diver inquired.

"Why, yon one wrecked just outside the bar," the fisherman answered. "Ye can walk about the seabed in that suit of yours?"

"I do it every day," the diver replied.

"Well, why don't ye walk out and get the treasure?" The diver smiled. "Show me the treasure, and I'll soon get it," he said. "Where is it?"

Solemnly the fisherman looked at the diver. "My father, he told me, and his grandfather, he told him. A mighty ship from Spain it was, full of treasure, that went down in a storm. They saw it from the shore here."

Puffing away at his pipe, the diver considered the matter. The story in his judgment might easily be true.

"Show me the spot, and we'll share the treasure, if there is any," he said.

"All right," the old fisherman agreed. "She's there all right. Sometimes we catch our gear in her."

Completing the task on which he was engaged, the diver began his search for the sunken treasure. Day after day he and the old fisherman went out in a rowing-boat, threw a grapnel over the stern and dragged it about the seabed in the hope of lighting on the wreck. Many of the villagers laughed at them and thought them crazy, but the two treasure-hunters paid no heed. They just went ahead with their monotonous task, buoyed up with the hope of the treasure to come.

The end of the first week saw them as far off the treasure as they had been on the first day. They dragged on through another week with a like result. A month of fruitless endeavour failed to rob them of their faith in the truth of the old story of the wreck. Week after week they searched the area in which the wreck was supposed to lie, tugging placidly at the oars, dragging the grapnel along the bottom.

One day the fisherman was rowing slowly along when the diver felt his grapnel catch in something. He gave the rope a sharp tug, then another, but the grapnel held firmly.

"We've got her," he said.

Marking the spot with a buoy, they rowed ashore for the diving suit and air-pump, then they went back to where the buoy floated on the surface. The diver donned his suit; the fisherman screwed the helmet securely into place, started to heave the handle of the air-pump as the diver went over the side and slid down the shot rope to the bottom. The ghost of the galleon greeted his eyes, the skeleton of the ship of long ago. For three centuries she had lain undisturbed in her watery grave, slowly rotting away until she had all but vanished. The diver climbed over the rotten remnants of the hulk into what had once been the hold of the ship. The place was full of weed; fish fled at the approach of the strange monster that was invading their domain; barnacles and sea-growth flourished on the decaying timbers.

With the same patience that had enabled him to locate the wreck, the diver searched the seabed until at last he came on what appeared to be several small barrels. He went up to them, tapped them. The much talked-of treasure was his at last. Beneath his fingers were solid stacks of Spanish doubloons, from which the wood had long since perished, leaving the coins still shaped like the barrels into which the Spaniards had packed them when they set out on that ill-fated expedition of theirs to conquer England.

These two men, with a diving suit and rowing-boat, found a greater treasure than has fallen to many a powerfully-equipped expedition, and it is strange to think that the fisherman who hauled the doubloons up from the bottom was probably a direct descendant of one of the Irish peasants who stood on the shore on that wild Armada night in 1588 and watched the mighty Spanish ship founder. The diver had the good sense to realize that there might be something in the old story, he spent weeks investigating it, and he reaped a snug little fortune as his reward. Nor did he squander the treasure that Fate flung his way. The same good sense which enabled him to find it also enabled him to keep it, for he turned his Spanish doubloons into a row of houses which he called "Dollar Row" in order to perpetuate his good luck.

Treasure hunters examining the bed of Tobermory Bay in the Isle of Mull through a special instrument invented for the purpose

Hard at work hunting the treasure of Tobermory. Washing the mud and sand dredged up prom the bay in order to find the Spanish doubloons reputed to be lost here over three centuries ago when wild weather helped drake to rout the Armada.

It is another tale of the Spanish Armada, a tale which up to the present has not ended quite so happily, that lures men to try their luck in the Bay of Tobermory in the Isle of Mull just off the west coast of Scotland. Somewhere beneath the waters of this pleasant bay is averred to lie a treasure so prodigious that it would make its discoverer a millionaire twice over. Here, if tradition speaks truly, a man has the chance of dragging from the seabed beautiful jewels and wonderful golden cups, with Spanish doubloons worth at least £2,000,000 which went down with the *Florencia.*

Many who have studied the question believe that the *Florencia* undoubtedly sank here, but an element of doubt creeps in when it is known that the Spaniards themselves swore that the *Florencia* returned after the disastrous expedition. During the Great War the British Government did its best to conceal the loss of H.M.S. *Audacious* in order to deceive the Germans as to the strength of our navy, and it may have been the Spaniards, three centuries ago, who introduced this practice. About this, nothing is known with certainty. It all happened a long time ago, and the years have tended to obscure the facts. Whether the statement that the *Florencia* returned was true, or whether it was a deliberate falsehood spread forth to give her enemies the impression that Spain was still strong in ships of the line, is an open question.

Whatever be the name of the vessel, the evidence that a Spanish galleon actually did founder in Tobermory Bay in 1588 seems fairly strong. Moreover, it is backed up by material facts in the shape of a cannon, some cannon balls, a weapon or two and a doubloon that have been brought up from the bottom of the bay by different treasure-hunters.

From what we can gather of that distant happening, it appears that the Spaniards, sailing down the Scottish coast in their galleon, and seeking perhaps to replenish their water-casks, must have made a foray or two ashore. During one of these they captured a Highland chief, one Donald Glas M'Lean, whom they held prisoner aboard their ship. So bitter a blow was it to the Scottish chieftain that, reckless of his own life, he sought a terrible revenge. Waiting his opportunity while the ship was anchored in Tobermory Bay, he managed to enter the powder magazine. In a moment or two his revenge was complete. The mighty galleon blew up and the proud chief accompanied her crew of nearly 500 Spaniards to their doom.

Many a tide has ebbed and flowed, many a storm arisen and subsided since that catastrophe. Timbers have decayed, and mud and sand have gradually covered up the remains. The treasure by now may be buried 20 or 30 feet at the bottom of the bay and, unless some lucky chance leads an expedition to hit on the exact spot, may remain buried there for ever. Divers may have walked over the treasure dozens of times without knowing that the gold and silver they were seeking lay actually under their feet.

The Duke of Argyll, who possesses the right to salve the treasure, has proved his belief in its existence by spending considerable sums in hunting for it. In addition he has given permission for several expeditions to prosecute the search, and these expeditions, in the aggregate, must have expended a deal of money. The lack of success on the part of previous expeditions seems in no wise to deter others from following in their steps, and the last expedition to work in Tobermory

Bay reflected the great changes of modern life by including a lady diver among its members.

Meanwhile the treasure of Tobermory Bay, which has excited the minds of treasure-hunters for many a generation, still awaits discovery.

## CHAPTER III

Whatever doubts there be about the treasure of Tobermory, there can be none about the treasure of the *Lutine*, for official records prove that when she came to grief she must have carried bullion worth over £1,000,000.

H.M.S. *Lutine* was a frigate of thirty-two guns, one of those wooden walls of Old England of which the poet sings. Not always had she sailed under the British flag. Time was when the tricolour of France broke at her masthead and French sailors crowded her decks, but Admiral Duncan captured her and brought her home as a prize, and thereafter it was the white ensign of England that flew at her peak and a captain of the British Navy who commanded her.

In the early days of October, 1799, at which time we were warring with Holland, H.M.S. *Lutine* was lying at Yarmouth, while the British troops garrisoned on the island of Texel off the Dutch coast were waiting anxiously for their pay. The *Lutine* was commissioned to carry the £140,000 due to the troops, and, hearing that she was departing for the Continent, many merchants sought permission to ship gold and silver by her for the relief of the merchants of Hamburg, who were financially embarrassed by the wars and the ensuing depression of the money market. The permission was readily granted, and 1000 bars of gold and 500 bars of silver were taken to Yarmouth and safely shipped aboard. In the ordinary course of business, the owners of the bullion went to Lloyd's and effected an insurance for the sum of £900,000.

On October of the year stated, the *Lutine* weighed anchor and sailed out of Yarmouth Roads on her voyage to Hamburg. As she bowled across the North Sea, the wind freshened and culminated that night in a terrific gale which the *Lutine*, gallant ship as she was, could not weather. The treacherous shoals off the Dutch coast reached out for her, and the mighty seas battered the life out of her and engulfed her. Of all aboard, but one human soul survived to tell of the wreck before he, too, succumbed from exhaustion.

The loss of the *Lutine* was a tremendous blow to Lloyd's. It meant that the underwriters had to find the sum of £900,000 with which to meet the claims of the insurers. Somehow they found the money and met all claims, thus adding fresh lustre to the name of Lloyd's and helping to raise it to the position it occupies to-day as the greatest and most powerful marine insurance association in the whole world. In return for their £900,000 the underwriters became possessed of the treasure—or rather the right to recover it! At that time, immediately after the calamity, when salvage operations naturally stood the best chance of success, the underwriters were prevented from doing anything at all owing to our war with Holland, and later on the Dutch Government made its position clear about the matter by claiming the wreck and all that was in it.

As the vessel lay, it was just possible to get to her when the sea was calm and the tides were at their lowest. It can be imagined that the Dutch fishermen made the most of their opportunities. Their government encouraged them by offering them one-third of everything they recovered, so the fishers found it profitable to leave their nets and spend their time fishing in the *Lutine*. Although the bulk of the treasure was beyond their reach, they managed during the next couple of years to lay their hands on a good deal of it. The Dutch Government received from the wreck treasure to the value of £56,000, and of this over £18,000 was paid to the salvors, while the rest was minted into Dutch money.

The amount of treasure which passed into the hands of the Netherlands Government during this period was not necessarily all the treasure that was taken out of the *Lutine*. It is possible, and indeed probable, that much of the treasure recovered was concealed by the fishermen salvors and used secretly to swell their own private hoards; but, even assuming that twice as much treasure was salved as was actually declared, there would still be a vast treasure worth over £1,000,000 remaining in the wreck.

A series of fierce storms wrought havoc with the wreck and placed her quite beyond the reach of the fishermen, who were at last forced to abandon their profitable quest. For years the wreck was the plaything of the storms, and not until Napoleon was safely imprisoned on St. Helena did any one give a thought to the treasure that lay amid the shifting sandbanks off the island of Vlieland. Then a Dutchman, going to his government, obtained a concession to salve the bullion on condition that half of what he recovered went to the government. For two or three years he fought the sea and sand to get at the treasure. No sight of gold or silver gladdened his eyes. Season after season, for eight years in all, he did his utmost to recover the fortune from the grasp of the sea, but without success. At last, weary of the incessant combat, he gave up the struggle and left the treasure to mock any other adventurer who might happen along.

The underwriters at Lloyd's, however, were not content to see the treasure which had cost them such a huge sum of money pass into the hands of a foreign nation, and at their request the British Government began to treat with that of Holland to induce them to relinquish their title in the wreck. The ways of diplomacy are often long and tedious, and this case was no exception. Many years elapsed before an agreement was arrived at and the Dutch gave up their claims and allowed the legal title in the treasure to pass to Lloyd's, its rightful owners.

For well over half a century the *Lutine* bore the brunt of the gales which afflict the Dutch coast, spending their strength on the belt of islands and the shifting sandbanks at the entrance to the Zuyder Zee. She was utterly lost amid the sands. Then came a terrific gale that blew for days, and the heaving waters washed the sand away from the wreck and made it possible to get at the treasure. For a period of five years, from 1857 to 1861, salvage men toiled away, and the result of their work was the recovery of bullion to the value of just over £40,000.

Once the salvors heaved the bell of the *Lutine* clear of the sea. It was brought to London and hung in the main hall at Lloyd's in the Royal Exchange. Whenever there is any important announcement to make to the underwriters about a ship being wrecked or an overdue boat reaching port, the bell of the *Lutine* is sounded to call the attention of all concerned.

Another time the salvors managed to bring up the rudder of the *Lutine*, and this was made into a chair and placed in the committee room at Lloyd's.

For another quarter of a century the sand and sea were left in undisputed possession of the wreck, then a new expedition set out to wrest the treasure from the encompassing sands. Right valiantly the salvors fought for that fortune, but luck was against them. Now and again they managed to bring up some of the coins that were lost in the *Lutine*, but the amount of treasure they recovered totalled considerably less than £1000 in all. So they discontinued further attempts and returned to England.

Since then more than one expedition has gone out to try to win the remaining treasure from the wreck of the *Lutine*. In the year 1908 the natives of Brightlingsea were astonished by the sight of a weird object that was anchored off the mouth of the river Colne. So strange a thing they had never seen before, and they puzzled their brains for an explanation of it. The curious object which caused so much amazement was a wonderful device for recovering the treasure of the *Lutine*. It was a great steel tube with a little iron ladder running down the inside of it. At one end were gigantic hooks for hooking it to the side of a salvage vessel, and at the other end was a steel chamber with a series of watertight compartments and air locks.

This marvellous contrivance, which took years to construct, was designed to be sunk in an upright position down to the wreck of the *Lutine*. It was equipped with water ballast tanks to sink it into place, and the steel chamber was furnished with cutting edges, so that the weight would enable it gradually to cut down through the sand until it reached the wreck.

Divers were to descend the iron ladder in the inside of the tube until they reached the submerged steel chamber. Then they were to enter the air locks where the water was kept back by compressed air, and walk out into the wreck. The divers would then communicate by telephone with the engineers in the steel chamber and direct the powerful pumps that were to suck away the sand until the treasure was reached. Once the treasure was found, the divers were merely to remove it to the steel chamber, whence it could be transferred to the salvage steamer above at their leisure. Excellent as the invention seemed, it did not recover the treasure of the *Lutine*.

Three years later, in 1911, another expedition more powerfully equipped than any of its predecessors resumed the search which had been going on for over a century. Notwithstanding the fact that the position of the *Lutine* was fairly well known, the obliteration of a landmark by a violent gale made it very difficult for the salvage men to find the wreck. The divers went down and searched the seabed vainly for a single sign of the old frigate. Not a spar was to be seen, not a rib of the hulk.

Captain Gardiner, who was in charge of the treasure-seekers, was a man of resource. He realized full well what had happened. The sand of the treacherous banks had completely buried the *Lutine*, and before he could make the slightest attempt to salve the treasure he would have to locate her and dig her out of her grave.

The problem of finding a wreck that lay buried deep in the silt would prove too much for any ordinary man, but Captain Gardiner was equal to the occasion. Among his equipment were some of the most powerful sand-pumps in existence, pumps capable of removing nearly a thousand tons of sand an hour. Dropping the end of one of these pumps to the seabed, he began sucking up the sand at a prodigious rate, cutting a deep channel right across the area in which the wreck lay. Slowly the pumps of the salvage ship devoured the sand and at last the salvors found the wreck buried 36 feet deep under a bank. The finding of the wreck was in itself a wonderful feat.

If only the other difficulties could have been overcome as easily, the treasure by now would have been won. But all the time the divers had to contend with the most difficult set of currents in the world. A strong tide, always running, plays incredible pranks with the bottom hereabouts. The submerged sandbanks are almost like cliffs some thirty feet high, and the tide moulds them and remoulds them almost day by day. A vessel at dawn may anchor in a deep channel, and by night the tides in one of their playful moods may have poured tons and tons of sand into the channel, completely filling it and building up a sandbank on the very spot where the channel existed only a few hours previously.

It will be realized how difficult this made salvage operations. The strong currents tended to wash the sand back directly it was removed, and the salvors were faced with what seemed like an endless struggle with the sea. They did not shirk the struggle; they went on dredging whenever the weather allowed, and they fought the tides most brilliantly by dumping the sand in such a position that it deflected the current right across the wreck. Thus there was a continual flow of water over the wreck to keep the site fairly clear and prevent the sand settling.

Meanwhile, they literally sifted the bed of the sea for traces of the elusive treasure. Every ton of sand sucked up by the pumps was poured through a gigantic sieve erected over the side of the salvage steamer. The sieve was like a giant birdcage, with a small mesh, and the men who watched the sand pouring through were more than once gladdened by the sight of a coin from the *Lutine*.

Seeking the treasure of the *Lutine*. One of the huge pumps sucking up the seabed at a prodigious rate and pouring it into the giant cages which sifted it for traces of the long-lost treasure.

They were weeks battling with the tides before the sand was cleared from inside the vessel and around the hull, but the day came at last when the divers went down to investigate the interior for the long-lost treasure. Every one aboard was keyed up to concert pitch. It seemed certain that the *Lutine's* treasure was to be lifted at last.

But the divers found the place in a sorry state. Much of the wooden hull had, of course, been preserved by the sand, but the magazine, in which the treasure lay, had collapsed, and there was practically a solid mass of iron five or six feet deep lying on top of the bars of gold and silver. When the magazine collapsed, hundreds of cannon balls had poured all over the place and these had been rusted together by the action of the water, locking up the treasure as securely as though it had been in a steel safe.

The only hope of the salvors lay in blasting this mass of rusted cannon balls to pieces and removing them bit by bit. In no other manner could the treasure be reached. Accordingly they set about their task, and little by little blew away the first layer. It was slow, tedious work, and all the time the salvors were harassed by the thought that the autumn gales might spring up and put an end to their operations, undoing in a single night work which had taken them months to accomplish.

Day by day they continued steadily with the blasting, and they had just succeeded in blowing away the second layer of rusted cannon balls when the dreaded gales came on. Further work was impossible, and sorrowfully the salvors left that exposed spot and went to Amsterdam to lay up for the winter.

A little more time, and they might have succeeded in their quest. There is evidence that they were somewhere near the gold, for one of the pieces of rust brought up bore the impression of a gold ingot, and when this rust was treated with acid it yielded five grains of the precious metal to prove that the gold was quite close.

Ten divers and a powerful plant had been seeking the *Lutine's* treasure for nine months. A small fortune had been spent on the operations. The workers removed a veritable mountain from the seabed, and they were rewarded with five grains of gold. They had shifted a million tons of sand to find five grains of gold! In this way does Fate taunt the deep-sea treasure-hunter.

The following winter the wreck was buried under 5 feet of sand by the tides, and by now she is lost once more, buried perhaps deeper than ever. The exposed position and the strong tides have kept the *Lutine's* treasure safe for over a century. But whether they will keep it safe for ever, no one can say.

It is a dozen years since I fingered one of the silver coins salved from the *Lutine*, and wondered whether the treasure was to be recovered at last. Still the *Lutine* is not forgotten, and only a few months ago I received from Lloyd's a letter from an inquirer in Vancouver who desired full details of the wreck, with a view to carrying on further salvage operations. I sent him the particulars he required, but so far I have not heard of operations being started.

For over a century wind and wave have beaten the men who sought to recover the wealth of gold and silver that went down with the *Lutine* on that wild October night. The fortune still lures men on to win it, and, in spite of the many disappointments, a lucky turn of the wind and tide, combined with improved salvage appliances, may yet make some future treasure-hunter a millionaire.

# CHAPTER IV

Without the diver, treasure-hunting beneath the waves would be impossible. The salvage expert may make the most brilliant plans, collect the most up-to-date and scientific plant to assist him, but in the end it is the diver who carries the work through, and upon the courage, determination and skill of the diver the success of the expedition depends. To dive to a depth of 5 fathoms, or 30 feet, is a task that the average man could accomplish without much difficulty; most men, too, would be able to reach a depth of 10 fathoms or 60 feet, if they were in decent physical condition. But at 15 and 20 fathoms and over the body is called upon to stand exceptional strains and so exceptional men are necessary.

Quite apart from the many risks, deep diving is very arduous, and seldom are men found with the physique that will enable them to dive 100 feet and over. The deep-sea diver must be trained like an athlete, perfectly sound in wind and limb and heart, and in tip-top physical condition. A fat diver stands little chance of attaining great depths, so the finest divers are generally on the slim side, men without an ounce of superfluous fat and with muscles tough as steel.

The physical strain placed on the body and heart merely by diving to these great depths is not generally realized. To ask the human body to undergo pressures three, four and five times greater than atmospheric pressure is expecting the body to undergo strains three, four and five times greater than the body was built to stand. It is like expecting a motor-car, designed for a load of 30 cwt., to carry a load of 6 tons. We should not expect the car to do that. Yet we not only call upon the human body to perform similar feats, but the body actually does perform them without collapsing.

The crack sea-diver is almost as difficult to find as the swimmer who can conquer the English Channel. When it comes to doing actual work at depths of 100 feet and over, the strain on the diver's body is indeed very much greater, for his exertions use up so much oxygen that his heart is called upon to pump at an increased speed in order to replace it. All the time, of course, the diver is breathing compressed air, thus the pressure of the sea on the outside of his body is practically counterbalanced by the pressure of the air inside his body. While the weight of the sea is trying to crush him inwards, the compressed air is pushing outwards, so the air pressure within equalizes the water pressure without, and the diver is enabled to work in perfect safety under a mass of water that would crush an unprotected man flat.

We might liken the water pressure to six men who are pushing hard against a door and striving to open it, while the air pressure resembles six men pushing against the other side of the door to keep it closed. With both teams equally matched in strength, the door remains quite unaffected by the contest if it be solidly built of oak. But if it is a weak door, the strain of the men pushing against it will probably break it.

Breathing compressed air not only places a strain on the lungs, but it tends to fill the body with an excess of nitrogen. This nitrogen may easily form tiny bubbles of gas, and these bubbles, if they reach the heart, might cause the death of the diver or bring on that dread paralysis known as diver's palsy, a disease which renders the lower part of the diver's body quite useless.

Strangely enough, it is not in going down that this danger threatens the diver, but only in coming up. If he comes up too suddenly, the excess of nitrogen in the blood bubbles like the tiny bubbles in a siphon of soda and at once his life is threatened. The bubbles are due to the pressure of the water on the outside of the body growing suddenly less than the pressure of air inside the body, consequently the nitrogen seeks to escape in bubbles just as the soda-water seeks to escape when the key of the siphon is depressed. The pressure inside the body cannot adjust itself quickly enough to the lessening pressure outside, and these bubbles are the result.

To avoid this risk, it is necessary for the diver working at great depths to come up very slowly. He may slide down the shot-rope to a depth of 120 feet in a few seconds, but, should he stay longer than half an hour at the bottom, he must not come up in less than fifty-seven minutes if he would avert danger. He may come up to 40 feet in eighty seconds, or at the rate of a foot a second. Then he must rest and exercise his legs and arms on the shot-rope for five minutes before ascending another 10 feet to the 30-foot level. Here he must rest for a further period of fifteen minutes, and do those exercises which help to rid his muscles of their excess of nitrogen. Ascending another 10 feet, which brings him to within 10 feet of the surface, he is compelled to rest for twenty-five minutes to allow the excess of nitrogen to pass from his blood, after which he may rise to the surface.

If a diver happened to remain an hour at a depth of 200 feet, he would have to spend four hours in coming to the surface to avoid any ill effects. The exceptional diver who is able to reach this depth should not, however, remain at the bottom for more than twelve minutes. This is the safe time, and he can then make the ascent to the surface in thirty-two minutes.

Remarkable diving experiments were carried out by the British Admiralty some years ago, during which naval divers attained the record depth of feet, a record that was long unbeaten. As a result of these experiments, tables were drawn up showing the time that a man might remain in safety at certain depths, and indicating the rates at which he could come to the surface and the depths at which he must rest to allow the pressure inside his body to adjust itself to the pressure of the water outside. These tables are followed the wide world over, and they have made diving one of the safest of occupations, despite the grave risks the diver is continually running.

Diving was, in fact, so dangerous that exceptional precautions had to be taken, with the result that the diver who walks about the bottom of the ocean to-day may be far safer than a man walking across Piccadilly Circus. The safety of the diver is most carefully watched over, but no one can foretell when a motor vehicle is going to run down some one crossing a busy road.

Never was knight attired for the tourney more carefully than the modern diver is clad before venturing into the depths. It is cold working at the bottom of the sea, and to guard against the cold the diver dons warm woollen sweaters and socks, sometimes wearing two or three sweaters and two or three pairs of thick socks. When he is dressed in his woollies, the diving dress is fastened about him just as the armour was fastened on the knights of old. There is a certain ritual about the performance which must be obeyed. First of all the shoulder pads are carefully tied on to take the weight of the head-dress, then an assistant helps him into the rubber diving dress and opens the tight cuffs for the diver to slip his hands through. The diver sits down while the assistant ties up the inner collar of the diving-dress and adjusts the various screws that are to secure his helmet. But before that is fastened into place the feet are slipped into the boots, each with its 16 lb. sole of lead.

Ever so carefully the diver's helmet is put on, for his life depends upon it being properly fastened. The air-pipe must be carried from the back of his helmet up under his arm to the front of his body where he can reach it easily and yet not find it in his way. The air-pumps and the valves in his helmet are most carefully tested to see that they are working properly. Then the diver gets on the ladder leading overboard and a lead weight weighing 40 lb. is adjusted across his breast and another similar weight is fastened over his back to enable him to sink to the bottom. The glass of his helmet is screwed up, the pump is set going, the diver waves his hand to indicate that all is in order, and the attendant after a final look round gives the diver a smart tap on the top of the helmet to inform him that he may go down.

Thenceforward the life of the diver is in the hands of the attendant, who never lets go of the lifeline and air-pipe until the diver comes to the surface again, feeling the diver at the end of the pipe just as an angler feels a fish at the end of a line, taking in the slack pipe to prevent it fouling rocks and wreckage, paying it out as the diver requires.

The coming of the submarine telephone has certainly lessened the risks of the diver, for he can now talk to the men in the boat and tell them what he wants and how he feels. If anything goes wrong and his lines become entangled, he can inform those at the surface, who can quickly send down another diver to assist him. In comparatively recent days it was necessary to signal by means of the lifeline and air-pipe, a certain number of pulls meaning certain things in accordance with a code in use by all divers. When a diver wished to convey a special message he had to signal for a slate to be sent down, and on the slate he would write what he wanted to convey. It was a slow and cumbersome method which has been rendered obsolete by the submarine telephone, which was invented by that famous submarine engineer, R. H. Davis, the head of Siebe, Gorman & Company.

For ages men have dived for sponges and pearls, remaining at most not more than a couple of minutes at the bottom. The ancients were fully alive to the advantages of an invention that would assist men to remain under water for considerable periods, and they were puzzling their heads about diving dresses centuries ago. These early inventions, however, were very crude, one being a sort of barrel with holes through which the arms could be passed, another a metal cylinder which covered the head down to the waist where it fitted into leather breeches. Very strange and wonderful they appear to modern eyes.

No less strange and decidedly more wonderful is the up-to-date diving dress which has grown out of the invention of Augustus Siebe in 1819. For eighteen years Siebe experimented with his first type of diving dress before he achieved, in 1837, the form of dress which is closely followed to-day. Various people have added improvements, but Siebe's form of dress is the one in common use, and the firm of Siebe, Gorman & Company which he founded to supply his diving dresses are to-day the greatest submarine engineers in the world.

Inventors have for long been concerned with the problem of a diving dress that will allow a diver to go to any depth without danger. The greatest risk of course, is that he will be crushed to death by the pressure of the water, and to overcome this danger more than one man has invented an all-metal diving dress with flexible joints. In appearance these diving dresses seem cumbersome, and the diver looks more than ever like a knight in armour.

Another form of dress largely in use enables the diver to descend in shallow water without relying on the usual air-pipe and pump. In such dresses the diver carries certain chemicals which not only purify the air he is breathing, but also furnish him with fresh oxygen. One chemical absorbs the poisonous carbonic acid gas given off by the breath, and the other chemical gives off fresh oxygen as the moisture of the breath touches it. The smoke helmet which enables men to enter a mine after a disaster, or a building full of foul fumes, is equipped with the same chemicals and made on the same principle as the diving dress. Instead of completely covering the man, however, this dress is made like a jacket reaching to the waist, where it is securely buckled.

In this dress it was impossible to penetrate the Redding pit, near Falkirk, from which five miners were marvellously rescued after being entombed for nine days, so several naval divers in regulation dress risked their lives in an effort to penetrate the workings to see if any other men still survived and to carry stimulants to them. Divers, at best, have the appearance of creatures from another world, and the effect of a diver, with his lamp, emerging from the inky water and coming suddenly on men who had been immured for a fortnight without food and were at their last gasp had to be carefully considered. Some of the survivors might have attacked him in their delirium and deprived their comrades of all chance of succour.

To avoid so untoward an incident, the leading diver carried with him a message for those men he hoped to find: "This is a diver come to save you. Don't touch him, as he cannot speak to you. We are driving a place for you. Don't sit down near the water, but keep clear of the damp. If any of your mates are far through, turn their heads downhill and that will help them until you are feeling stronger. The diver cannot come up the hill out of the water to help you, because his tools are too heavy. He will come back regularly and feed you. You must not drink more than half a cupful of beef tea each. Wait and take a rest before you drink another half-cupful. On this paper write who you are. You will be got out soon."

Alas, for human endeavour, that message never reached

the poor fellows for whom it was intended! The great falls of roof choked the roads and proved an insurmountable barrier. Raging, but exhausted, the divers had to bow their heads in defeat.

So commonplace is the diving dress that it no longer excites curiosity. Yet it remains one of the wonders of modern civilization. Merely by utilizing the sap of a tree, which we know as rubber, and fresh air, men are now able to work and live at the bottom of the sea.

# CHAPTER V

It was in 1891 that the steamship *Skyro* pulled out of the port of Cartagena, in southern Spain, and set her course for London. The coast of Spain glided by as she proceeded through the blue seas of the Mediterranean, speaking Gibraltar as she passed, and setting her nose north to skirt the coast of Portugal. Oporto dropped far astern, and the Portuguese coast changed to the western coast of Spain as a fog quietly stole down and blanketed everything. The fog was dense. Not a thing could be seen, and the warning notes of the *Skyro's* siren blared monotonously as she felt her way blindly along. The captain and officers stared anxiously ahead, hoping that the fog would lift; but there was no sign of a break around them, nothing but fog and the sound of their siren to warn passing ships.

Of a sudden the ship staggered and halted. It was as though a giant hand had reached up from the depths of the sea and grasped her keel. The crew were thrown higgledy-piggledy. There was an awful rending sound as the *Skyro* swung onward. She had struck the dreaded Mexiddo reef off Cape Finisterre, and as she slid over the cruel rocks they literally tore the bottom out of her. Slowly she carried on, while that rending sound continued, and twenty minutes after striking she slipped off the reef and plunged to the bottom.

A few hours later the bell of the *Lutine* in the Royal Exchange was clanging loudly. The underwriters paused in their work. All voices were stilled, and the scarlet-coated crier, mounting his rostrum, announced in stentorian voice that the steamship *Skyro* had struck the Mexiddo reef off Cape Finisterre and was a total loss.

Then the bustle of business began again, but a little knot of underwriters gathered together and started to talk quietly. They were interested in the silver bars that the *Skyro* carried.

"What about salvage?" one inquired.

Another, who joined the group, shook his head.

"Hopeless. She's down in 25 fathoms, or more."

"You never know," said one man who was more intimately concerned.

He was quite right. You never know. Men manage sometimes to achieve the impossible.

Fuller information made the salvage seem more remote than ever, for instead of being down in 25 fathoms, as had been supposed, she was several fathoms deeper, and her keel, resting on the bottom, must have been well over 30 fathoms from the surface. Nothing had ever been salved from such a depth before, and it seemed unlikely that any man could go to this depth and survive the enormous pressure.

However, an expedition went out and fought to get at the treasure, but the depth was too great, and at last the salvors withdrew from the spot. Four years passed and there came to the underwriters another offer to attempt to salve the silver. The salvage vessel anchored off the fringe of the reef that had stripped the bottom out of the *Skyro*, and the diver slid down the shot-rope to try to find out how the wreck was lying and if possible to bring out the precious bars. Before he could do anything of importance, however, bad weather set in and drove the salvors back to harbour. But the lesson learned from that attempt was that, if the treasure were to be recovered, more powerful diving gear would have to be used.

The winter months were spent in obtaining much more powerful gear from England, and the following season, directly the fine weather set in, the treasure-hunters repaired to the Mexiddo reef to try once more to achieve the impossible. The diver feared nothing. Brave as a lion, he took the shot-rope in his hands and slid straight down to the deck of the *Skyro*, which was 171 feet below the surface. Carefully and quietly he surveyed the ship, seeking the cabin in which the silver was stored. The deck had collapsed on top of it, and the only way of getting to the treasure was through the deck.

Angel Erostarbe, the diver, came to the surface and reported what he had seen. Difficult as was the task, it seemed to him by no means impossible. So he dropped down the shot-rope again and again. Gradually and with infinite patience he blasted away the deck, fixing his charges and withdrawing while they exploded.

So exposed was the wreck that at times he could hardly keep his feet. Time after time dirty weather came and prevented him from working at all. The difficulties left him unmoved. He set his teeth and stuck to his task. He was working at a record depth, a depth which most experts considered was beyond the reach of a diver at all. The diver did not worry about this. All he thought about was getting at the treasure.

To attain his end he practically blew the ship to pieces, and his marvellous feats of endurance were crowned by the recovery, in two seasons, of fifty-nine bars of silver worth £10,000. It was a stupendous feat which has never been equalled since. At times he was actually working in 183 feet of water, so it will be seen that he was an exceptional man. Toiling at this depth—where his body was subjected to the huge pressure of about 95 lb. to the square inch—left its mark on him, and he was never the same man again. His share of the treasure amounted to £500.

Compared with this, the recovery of the treasure from the *Oceana*, when she was sunk in the Channel in 1912 as the result of a collision, was a comparatively simple matter, yet it was not without its difficulties. The *Oceana* went down in 90 feet of water and only her masts peeped above the surface when the salvors arrived on the spot. Plans of the ship were obtained from the owners and carefully studied so that once the divers got aboard they would know exactly which way to go.

It is difficult enough for the average man to find his way about a strange liner when she is afloat, so it can be imagined how difficult it must be for a diver to wander about such a vessel when she is 90 feet under water. All the time he is adventuring through the saloons and other compartments, he is running continual danger of his air-pipe catching on something and tying him up. He may lose himself. Doors may slam to with the current and imprison him while cutting off his air supply. The men manning the air-pumps will quickly find out that something is wrong, but by the time assistance is sent the imprisoned diver may easily be in a sorry state.

The ordinary difficulties were intensified in the case of the *Oceana* by the strong currents racing down the Channel. So strong were they that even in favourable weather it was only possible for the divers to work for one hour a day when the tide was at its lowest. To make matters worse, there was so much sand in suspension that the divers could see nothing at all. The electric lamps which it was hoped would help them were quite useless. The divers were like blind men, groping in the dark, feeling their way about the ship and working by touch alone.

They blasted their way through two decks and, stumbling along a passage, found the strong room. Ingot by ingot, they took out the treasure and sent it to the surface, where each bar was carefully checked and marked off in the records as it was recovered. If only all the treasure had been carried in the strong room, the game of blind man's buff on the part of the divers would have been at an end. But a good deal of the silver was stowed in the after hold, and before the divers could get at it they had to force their way through three decks. Ultimately all the treasure, to the value of £700,000, that went down in the *Oceana* was recovered and the treasure-hunters sailed away in triumph with their spoil.

The astonishing feat of Erostarbe was almost equalled by Alexander Lambert, one of the finest submarine workers who ever lived and the chief diver of Siebe, Gorman & Company. He covered himself with glory during the building of the Severn tunnel when, owing to an error, a door was left open and the workings were flooded. The water rose some forty feet up the shaft leading to the workings, and it was impossible to continue building the tunnel until this door was closed.

Realizing that the only thing to be done was to send down a diver to close the door, the engineers called on Lambert to essay the task. Descending the ladder of the shaft, Lambert disappeared under water and made his way to the bottom, where not a single ray of light could penetrate. Feeling round the wall of the shaft, he found the opening to the tunnel, and began slowly to venture along. But the rush of water had worked tremendous havoc, and the tunnel was strewn with debris which was most difficult to negotiate. At any moment Lambert's air-pipe was in danger of being cut by some projecting piece of the wreckage, and, in addition to the weight of his dress, he was terribly hampered by the weight of the 1200 feet of air-pipe which he was forced to drag along after him as he stumbled about the workings.

The wreck of the *Oceana* which sank in the English Channel as the result of a collision. She had treasure aboard worth £700,000.

Divers going down after the treasure of the *Oceana*. Note the double-handed air pump which the two attendants are working.

Hearing of Lambert's baffling problems, Fluess, the inventor of the diving dress which dispensed with the air-pipe, volunteered to go down in his self-contained dress and see what he could do. Fluess was a clever inventor, but the only diving he had ever done was in connection with his experiments on his new type of dress. Besides being a clever inventor, he proved himself a man of courage.

He arrived on the spot with his diving dress, and studied the plans of the workings to find out which way he had to turn when he got to the bottom of the shaft. He thought it would then be just a question of walking through the tunnel, finding the door and closing it, little knowing that the place was in a deplorable condition and beset with all sorts of obstacles.

"Lambert had better go down first to take off my life-line and tell me which way to go. He knows the place a bit by now," the inventor suggested.

Accordingly Lambert went down and waited 40 feet under water in the inky blackness for the inventor. Fluess made his way down the ladder in the centre of the shaft, taking a firm hold of the rungs with his hands and feeling for the next one with his foot. As it happened, the ladder was short of the bottom by some 10 feet, and they had forgotten to inform him of this fact. Fluess, coming to the end, felt as usual for the next rung. It was not there, so he lowered himself one rung by his hands, expecting to touch the bottom with his feet. His feet merely churned in the dank water, so he went down rung by rung until he was clinging to the last rung with his hands. After vainly feeling with his feet for the bottom, he let go his hold and dropped about 6 feet.

Some boards creaked and tipped ominously under him as he landed, then he felt his way round until he came to Lambert. The diver took off the inventor's life-line, and Fluess fared forth into those underground workings some 200 feet beneath the surface of the green fields above. It was a weird experience. At first he tried to walk, and being without any guide whatsoever he lost all sense of direction. Then he tried for the sides of the tunnel, but there were ditches and wreckage which brought him down so often that he was forced back to the centre of the road. So he went down on his hands and knees and began to crawl along, feeling the sleepers of the tram-track with his hands, using them as a guide. He came, after many tribulations, to a place where the sides and roof had fallen badly and very laboriously managed to crawl over the heap of debris. After struggling about the underground tunnel for an hour, he was forced at length to turn back. Another and yet another attempt he made, each time getting a little farther along the tunnel.

"Why not let me try?" said Lambert at last.

"Very well," said the inventor.

Lambert had never before used the new type of diving dress, but that did not deter him. He got into it and had a short trial dive one afternoon, and the next morning went down the shaft to try in dead earnest to close the sluice which was letting in the water.

The inventor went down too, and sat there waiting, waiting, and wondering what had happened to Lambert, and whether the new diving dress was going to justify his hopes. The diver, meanwhile, was fighting his way forward over the numerous obstacles in the tunnel, crawling over the falls and squeezing between the roof and the debris. It was nervy, risky work, for he did not know whether another fall would come and bury him or close the small exit, nor did he know whether he could manage to find his way back again. Under such difficult conditions, anything is possible.

Nevertheless, he managed to get to the door that had caused all the trouble. Feeling round, he found one of the valves open and succeeded in closing it. Then he investigated the door and found that before he could close it he would have to take up a couple of rails that were obstructing the entrance. Away down in the bowels of the earth in that flooded tunnel, far from help, relying upon his own strength and courage alone, he struggled with the rails and managed to get one free. The other baffled all his efforts, and reluctantly he turned round and made his slow way out of the tunnel, after being away for an hour and a half.

He was drawn up with Fluess, and directly their helmets were unscrewed the inventor turned to Lambert.

"How far did you get?" he asked.

"Right up to the door," said Lambert. "It's wedged open by two rails. I managed to get one away, and to close one of the valves. I think, if I take a crowbar along, I shall be able to manage it all right."

Sure enough, he went down and fought his way along the flooded tunnel again. After a struggle, he levered the other rail up and succeeded in passing beyond the door to close another valve, afterwards shutting the door that had caused all the trouble. Before returning, he knew that one more valve must be screwed up to keep the water back. The tips of his fingers slid over the surface of the door like those of a blind man until he found the valve, then he screwed it round until it would screw no more.

He little knew, as he screwed away, that he was screwing the valve open, but so it was. That valve, instead of following the usual rule and screwing up to the right, actually screwed up to the left. Whether any one knew of this variation, or whether the engineers forgot it in their fight to free the tunnel of water, the fact remains that no one told Lambert, who unconsciously screwed the valve open, with the result that the tunnel took longer to pump out, because the water still poured through this valve. Not until the water was overcome was the mystery of the open valve solved.

The diver who performed this brilliant feat salved many fortunes from the seabed, and was perhaps the greatest hunter of sunken treasure who ever struggled into a diving dress. Even the experts, however, thought little of his chances when he went out to try to salve the treasure of the *Alphonso XII.*, which was down in 160 feet of water off Point Gando in the Grand Canary.

"Lambert has the job in hand," said one.

"He can't do it. She's too deep for mortal man to tackle!" came the reply.

Lambert dropped down to the deck of the *Alphonso*, and knew that a fortune lay under his feet. He paced the deck until he came to the exact spot beneath which the treasure should lie. Then he began to investigate the ship, but, skilled as he was, he would not face the risk of getting lost in its interior, of fouling his lines while he groped his way in the darkness along passages and through cabins and saloons to the strong room. To venture into the bowels of the ship would probably mean that he was going to his death.

He summed up the situation. The treasure lay beneath two decks. To tear a way through with crowbars or to chop a way through with axes was impossible. Every movement at that depth was terribly exhausting, and he had to rest, in order to recover, after doing the slightest thing. His only means of getting the treasure was to blast a way through with explosives, to harness explosives to do the work and thus save his own energy.

He set to work and after tremendous trouble blew through the top deck. Clearing the shattered pieces away, he let himself down into the saloon, and began his attack on the second deck. It, too, succumbed to the mighty concussions of the explosives, and Lambert dropped into another saloon. He looked about him, and in the floor at the farther end he found the entrance to the strong room. The trap-door resisted his efforts, but in the end Lambert's crowbar, skilfully wielded, prised it up.

Lambert went into the treasure-room and saw the little chests of treasure, each one of which contained a fortune. He signalled to the surface, and a cable was let down. The tremendous pressure hampered his movements, made them seem slow and clumsy. Nevertheless, he raised a chest full of treasure and managed to slip a rope beneath it, then he secured it to the hook hanging beside him. The signal was given, and Lambert watched his first haul of the treasure mount through the opening he had blasted in the ship. That chest swinging on the end of the rope was full of gold coin worth £10,000!

Every time he braved the depths to seek the treasure he took his life in his hand, but he did what he set out to do, and in the end he managed to send to the surface seven boxes of treasure worth £70,000, leaving another two boxes worth £20,000 to be recovered at a later date. Lambert received £3500 as his share in this deep-sea enterprise, in addition to his pay of £40 a month and all found.

Thrilling as were these treasure hunts, the most romantic story of all is that of the *Hamilla Mitchell.* Here we have treasure and pirates and a desperate chase all mixed up in the most approved adventure-story style. Only, unlike a work of fiction, this story happens to be true.

The *Hamilla Mitchell* came to grief on the Leuconna Rock, near Shanghai, and carried down with her £50,000 of specie. She was a total loss, and the underwriters, after paying the insurance, considered the question of trying to salve the treasure. They instructed an expert to visit the scene and report on the case. The expert in due course considered that the case was hopeless, that the specie was lost for all time, and that the wreck had gone down in such deep water in so exposed a position that it was much too dangerous for divers to work there—not a very cheerful report for the underwriters to receive.

There, for a time, the matter rested. Then upon the scene came a Captain Lodge with an offer to do his best to recover the treasure. The underwriters, unwilling to allow the specie of which they were the owners to remain at the bottom of the sea, agreed gladly to the proposal that was placed before them. Captain Lodge considered the problem most profoundly. He knew that what was lost would not be won back easily, that the odds were, indeed, very much against a single ounce of the precious metal ever again seeing the light of day. This did not dismay him. Securing the services of two clever divers, named Ridyard and Penk, he made the trip to Shanghai, taking out with him some special diving apparatus—the finest and most powerful equipment to be found in the world.

He wandered about Shanghai looking for a vessel that would suit his purpose, and, coming across a small sailing craft, chartered her and proceeded on his quest for the wreck. Small as was the salvage vessel, she was yet too large to take inshore among the high rocks, and so the divers had to prosecute their search from the small boat which they towed behind. They searched here, they searched there, dropping over the side of the boat in their cumbersome dress, facing all the unknown perils of the unknown depths. Now they were carefully exploring a ledge perhaps only 20 feet deep, and a little later they would be slipping down the face of a chasm that plunged sheer into the sea for another 100 feet or more. They did not spare themselves in that search, for at times they penetrated to a depth of 160 feet.

They were investigating a ledge one day when a dark mass loomed up at one end. They approached it, to find the wreck at last, noting with satisfaction that it was in a comparatively shallow depth which made the prospect of salvage fairly easy. Their jubilation was cut short, however, as they drew nigh. It was the stern that held the treasure, and the stern was missing!

Fate had once more been up to her tricks. The *Hamilla Mitchell* had settled with her stern overhanging deep water. Not for long did she remain intact, for the gales soon broke off the unsupported after end, which slipped off the ledge into the abyss, where the divers managed to locate it in 156 feet of water.

The never-ending lines of bubbles from their outlet valves flowed upward to the surface as they slowly explored the stern and prepared for their assault on the treasure-room. It was a most dangerous as well as a most difficult task to work in that treacherous chasm. The currents were strong, the rocks were sharp, and the possibilities of air lines being cut or fatally fouled were not pleasant to dwell upon. Nevertheless, they stuck to their task and eventually Ridyard managed to break a way into the strong room.

The sight which met his eyes as he gazed through the windows of his copper helmet was like a scene from some fairy tale. The light, filtering through to that great depth, enveloped the hold in a sort of twilight gloom, and all over the place he dimly saw heaps of dollars scattered about. He stooped down to the treasure chests, to find that woodboring worms had eaten many of them quite away and the contents of the boxes were spilled in all directions. He walked about on a floor of solid gold; golden coins slipped about under his leaden soles.

Anything more romantic would not be easy to find, yet the romance did not appeal to Ridyard. He was working against time, knowing that he would not be able to stand the pressure for long. Every movement was slow and difficult. The water was striving to crush him; he was being saved from this terrible fate solely by the continual flow of air coming down the rubber pipe to his helmet.

Four times Ridyard underwent that ordeal of getting into the treasure-room and working under the enormous pressure until he was quite exhausted. On the last occasion he surpassed his previous feats of endurance and struggled doggedly on, loading up the treasure and watching it disappear towards the surface until he had sent up the contents of sixty-four boxes.

Strong and fit as he was, he became thoroughly worn out with the toil, so he signalled to those above and made his way slowly to the surface. They dragged him to the deck of the salvage craft and unscrewed his helmet. His face was lined, his eyes were very tired, and his body clamoured for moisture, although he had been immersed in it for a long time. Not a glance did he give to the treasure lying about, the fortune at his feet did not interest him.

"Give me a drink," he said. "I'm dying for a drink of water."

Penk nipped up a bucket and made his way to a spring at the top of the island under which they were working. Putting down his bucket to fill, he scanned the horizon, as sailormen will. A sudden amazement came over him. The sea was dotted with sails, all making in the direction of the island.

Wasting no time, he picked up his precious pail of water and ran down to the ship.

"What's up?" asked Captain Lodge, as Ridyard took his

much-wanted drink.

"The sea's full of junks, hundreds of them," Penk replied.

Taking his glasses, Captain Lodge quickly identified the oncoming ships as the junks of Chinese pirates who were making their way towards the island from the farther side to avoid being seen. There was no doubt in his mind as to what they were after. There was but one thing in that quarter worth having, and that was the treasure stored in the salvage craft. It was obvious that the pirates had been watching operations carefully. They had undoubtedly planned to allow the divers to recover the treasure, then they purposed stealing down upon the expedition unawares, wiping it out and looting the gold.

The pirates were in overwhelming numbers, and Captain Lodge realized instantly that the only thing to do was to run for it. Slipping the anchor to save the time required to haul it up, the salvors hoisted sail. Gradually they gathered way and stole from under the cover of the island. Directly the salvage craft appeared in the open, the junks altered course and started to pursue her.

Pity the poor salvors! The wind had practically failed them, yet they could see some of the junks bending to a lucky breeze and overhauling them. In desperation they put out the big sweeps and toiled like galley-slaves to force their craft through the water. Ridyard, tired as he was, took his turn at the oars to try to save the treasure he had salved at such risk. So the salvage boat crept along, with the pirates slowly gaining.

More exciting grew the chase. With anxious eyes the salvors watched the distance between their own craft and the Chinese junks growing gradually less. Harder than ever they strained at the oars, dipping them into the sea, throwing all their weight upon them, pulling until the muscles of their arms ached and their backs were nearly breaking.

It looked as though the salvors would lose their lives as well as their treasure when the sails, which had been flapping idly, began to swell. A puff of wind stirred their flag, and a steady breeze began to blow. It was none too soon. The salvage craft started to gather way again and forge through the water. Still the junks hung on. They were not going to relinquish their prize without an effort.

The pirates continued to chase the salvage craft right until sundown, when a friendly darkness hid pursued from pursuers and enabled Captain Lodge to shake off and lose the bloodthirsty Chinese pirates. In the end he managed to make Shanghai in safety with the rich treasure of £40,000 aboard, thus bringing to a happy ending one of the most exciting treasure-hunts ever known.

If Ridyard had not worked quite so hard and grown quite so thirsty, and if Penk had not gone to fetch that pail of water, the salvors would have remained in ignorance of the approaching pirates and would have met a tragic death at their hands.

That lucky drink of water saved a fortune of £40,000.

# CHAPTER VI

For months at a time during the past few years, a little ship may have been seen floating around a particular spot just off the coast of Donegal. Barges lay in her vicinity, barges laden with incredible tangles of pipes and cables. Boats pulled around from barge to ship, and fussy little launches came from the coast, remained an hour or two, and then departed. Occasionally a grim, grey destroyer glided up, moored for a time, and then steamed away. But the little ship remained, and strangers in those parts wondered what she was doing there.

That little ship was the salvage vessel *Racer*, engaged in the greatest treasure-hunt of modern times. Never before has there been such a treasure-hunt, for it was a national treasure-hunt, carried out on behalf of the British people by the British Navy, and backed by the whole power of the nation.

When the White Star liner *Laurentic* left the shores of England in January, 1917, she carried in her strong-room gold and silver ingots to the value of about £5,000,000 to settle some of Great Britain's bills for the munitions that were pouring out of the factories in the United States. The Treasury was naturally anxious for the specie to reach its destination as quickly as possible, for that £5,000,000 was destined for the pay envelopes of thousands of American factory hands.

Many a time the *Laurentic* had made the passage with saloons brilliantly lighted and crowded with wealthy passengers, but never before had she borne so much wealth as on this occasion. The advent of war led to her conversion into an armed liner, and those aboard were now fighting for the freedom of the seas and civilization.

Northward she steamed through the Irish Sea and at last began to breast the open Atlantic and point westward to New York. Malin Head, on the north coast of Ireland, loomed up and began to drop astern, and just when it seemed that all would be well came the blow that sent her to her doom. A violent explosion shook her, made her lurch and shiver, and many gallant fellows, watchful at their posts, were instantly killed; many more were trapped and drowned by the rush of water into the ship.

The survivors sprang to their emergency posts, while the wireless operator sent out a call for help. The captain realized that the *Laurentic's* days were numbered. Nothing could save her. The water poured through the rent in her side. More and more she heeled as the water gained. For a moment her bows lifted clear of the sea, then she disappeared in a swirl of foam, and the waves were strewn with wreckage and bobbing heads. When the tragedy was over, and the roll called, it was found that, of 475 officers and men aboard, 354 had gone to their last long rest.

The loss of life, the destruction of the ship, the sinking of the treasure, all were bitter blows. The gallant sailors were beyond recall, the ship was sunk for ever. As for the treasure, it was down in 120 feet of water, on a coast so fully exposed to the Atlantic gales that its recovery was an open question.

Prospecting for gold in the desert places of the earth has its difficulties and its disappointments, but what are these compared with the problems that confront the men who seek to wrest from the mighty ocean the gold it has swallowed? Unexpected dangers often confront those who seek the precious metals in the wild places of the earth, but the dangers of the diver are continuous. He trusts his life to a frail rubber pipe and a rubber suit, and directly the metal helmet is screwed round his neck, and he sinks into the depths, death starts to stalk him and does not give up the chase until the diver is once more aboard the salvage ship.

Some of the finest divers in the British Navy were told off for the treasure-hunt. They were eventually placed under the command of Commander Damant, who had played so important a part in the diving experiments carried out by the Admiralty a few years ago, and who had himself attained the record depth of feet in August, 1906. The fact that the cleverest diving expert in the British Navy was detailed for the operation is proof that the Admiralty realized that the recovery of the treasure would prove no easy task. No one knew at the moment exactly how strenuous the fight was going to be.

The first salvage craft, which was later replaced by the *Racer*, went off to the Donegal coast and swept the area in which the *Laurentic* had disappeared. The salvors found the wreck in due course, and they had the satisfaction of knowing that they were within 120 feet of a stupendous fortune of about £5,000,000. A bare depth of 120 feet of water separated them from the greatest treasure-trove of modern times, but the treasure could not have been more secure had it been resting beneath 120 feet of solid steel. Indeed, had the treasure been so buried, instead of underneath 120 feet of water, it would probably have been recovered very much sooner.

Despite difficult conditions, a certain optimism prevailed that the treasure would soon be brought to the surface. But the optimists reckoned without the enemy. Somehow the Germans managed to find out where the *Laurentic* was wrecked, and their submarines quietly waited their opportunity and began to make things hot for those engaged in the treasure-hunt.

One enemy submarine haunting the vicinity discreetly vanished as a British torpedo boat came on the scene. A day or two passed, and the torpedo boat was called for urgent duty elsewhere. Meantime, there had not been the slightest sign of the enemy underwater craft, which had apparently recognized that that particular spot was rather unhealthy and therefore to be avoided.

Feeling fairly secure, the salvors, according to an unofficial report, determined to get on with their job. A diver donned his dress, his helmet was screwed on, and the air-pumps began to heave as he dropped down to resume operations. He had been down but a short time when he felt himself plucked off his feet by a mighty pull on his life-line and air-pipe. He struggled to right himself, but it was quite useless. An irresistible force dragged him upwards; then he felt himself being drawn through the sea like a salmon at the end of a line.

Something was running away with him. It was an awful experience. He wondered what had happened and how it would end. His senses began to reel; he found a difficulty in breathing.

Somehow he managed to keep his head and act as the emergency demanded, closing the valve by which the air escaped from his helmet. A minute later he broke the surface.

He could hear the seas slapping the top of his helmet as he was dragged along at a smart pace. His heart pounded, a terrible humming droned in his ears, but he strove hard to retain his senses.

"What's up?" he thought. "What on earth's happening?"

He had no chance of finding out. He was prisoner in a metal helmet and a rubber suit. He knew he was at the surface, because of the light that filtered through the glass of his helmet and the seas that swished against the copper. As he was dragged along, he had a tendency to spin at the end of his line, which gave him a dreadful sensation.

In a dazed sort of way the diver was wondering how long the ordeal would last, when he suddenly felt himself plucked clear of the water. The next thing he remembers is something scorching his throat and the cool air playing about his head. He looked round and found he was lying on the deck of the salvage vessel, and he thanked his lucky star that all was well. Then he was placed in the recompression chamber aboard, so that the dangers of being dragged hastily from such a depth might be avoided, and the risk of bubbles of nitrogen forming in the blood averted. The air-pumps were set going to raise the pressure of the air in the steel chamber to the same pressure as that at which the diver had been working, and gradually the pressure was reduced until it was the normal atmospheric pressure and the diver was able to be taken out.

While he was on the bottom, a German submarine had stealthily approached the salvage vessel. Suddenly it started to attack, and the salvage steamer had to cut and run for it, dragging the unfortunate diver in its wake. The attack was so unexpected that there was no time to pull up the diver in accordance with the rules. To pull him up in the ordinary way would, as a matter of fact, have taken half an hour. There was no alternative but to tow him along willy-nilly and haul him aboard as they fled. The experience might easily have cost the diver his life, but the recompression chamber fortunately saved him from any ill effects.

After this rather exciting episode, it was decided that operations to recover the treasure would have to be postponed until more peaceful times. The treasure-seekers had their hands full in fighting the stormy seas and powerful currents, not to mention the great depth of water, without having to fight the foe as well.

At the end of the war, the battle with wind and wave for the treasure of the *Laurentic* was once more resumed. So exposed was her position that for fully half the year it was impossible for divers to work there at all owing to the storms that raged. Even in fine weather there were the currents to fight against. And their strength at times was almost incredible. They could swirl big boulders along the seabed as though they were but pebbles.

More than one diver, during his career, has experienced the sensation of being picked up like a feather and dropped over the side of the wreck on which he has been working. He might weigh roughly 160 lb. Slung over his back would be a 40-lb. weight, across his chest would be a similar weight, while each boot would be loaded with a leaden sole weighing 16 lb. Fully equipped he would turn the scale at about 3 cwt., yet the current has simply played with him as though he were thistledown. Its strength has been such that he could not fight against it. Consequently, he has been compelled to give up all ideas of work and return to the surface. It is indicative of what the salvors of the *Laurentic* had to contend with in this respect.

Two years at the bottom of the Atlantic had wrought a tremendous change in the once-proud liner. The divers found her plates corroded with rust, girders collapsing everywhere. The sheer weight of the water above her was crushing her flat, squeezing her into a shapeless mass just as you might crush a lily in your hand. Moreover, she was full of silt and mud. Strange fishes glided about her inky depths. Dread conger eels of mighty girth lurked in the labyrinths of the wreck.

In spite of the terrible condition to which the wreck had been reduced, the divers finally managed to locate the strong-room. The bubbles from their helmeted heads flowed ceaselessly upward as the exhaust air ascended to the surface. Slowly they made their way forward towards some bars, dimly seen within the recesses of the ship. They were in the treasure-room. The gold and silver lay about them. Some of the precious ingots barely peeped out of the silt.

The attendant on the salvage ship heard the telephone buzz.

"Hallo!" he said.

"We've found the treasure," said a voice from under the sea. It was a squeaky voice, for, strangely enough, talking in compressed air gives the voice a high pitch, and at this depth it would be impossible for a diver to whistle. The pressure of the air on his lips would prevent him.

No time was lost in lowering cables, and one by one the ingots began to speed to the surface. Then, all too quickly, the signal was given for the divers to ascend, and the treasure had to be left for another day.

That season ingots valued at £500,000 were recovered from the strong-room, after superhuman labour on the part of all concerned. So extremely arduous were the conditions that our crack divers could only work two spells of fifteen minutes' duration each day. Half an hour's toil beneath the sea took as much out of them as the ordinary day's work takes out of the ordinary man.

Once more the winter gales played havoc with the wreck, and next spring the divers found that the treasure was lost under a mass of twisted plates and girders. Imagine a street of lofty houses, then imagine that all the buildings were pushed suddenly down into the centre of the road, and you will arrive at some faint idea of what the ship looked like. Great girders were bent into all sorts of strange shapes; iron bars thick as a man's wrist were twisted into fantastic curves.

The only way to get to the treasure now was to blast a passage with explosives. The difficulties of the task were increased by the necessity of hoisting every bit of plate out of the wreck and towing it some distance before dumping it, in order to make quite certain that the plate would not again obstruct the divers. The placing of the charges in the most effective spots, and the withdrawal of the divers while contacts were made and the charges exploded, took a long time and entailed endless trouble. But the salvors kept at it doggedly, and bit by bit they cut away obstructing plates and girders weighing about 300 tons.

A diver going down to blow up part of a wreck to get at the treasure. The charge of explosive, weighing 50 lbs., is contained in the long tin over the side of the boat. Sometimes the explosive is packed in a canvas bag three or four feet long and three or four inches across.

Thus they opened up a way to the treasure, and once more began to send ingots of the precious metal to the surface. Things began to look rosy, and there seemed the prospect of making a clean sweep of all the bullion, when a terrific storm arose and stopped operations. When the divers went down again they found that more plates had folded down over the treasure, as if deliberately to prevent its abstraction. It was a dreadful disappointment, for very soon afterwards the autumnal gales put an end to the hunt for the season.

The next year the *Racer* was back again off the Donegal coast, eager to resume the great treasure-hunt. But it proved a terrible season. The weather seemed to mock the hunters. For weeks at a time work was impossible. As soon as one storm abated, another sprang up.

Waiting with all the patience they could muster, the divers at length got a chance of going down to the wreck. What a change the gales had wrought! No longer did the wreck bear any resemblance to a ship. She was just a great mound of twisted metal, partially buried in the silt. Plates and wreckage lay scattered over the seabed in all directions, covering an acre or two of space.

Once more the dangerous task of blowing away obstructions was resumed. Carried out as expeditiously as possible, it yet proved all too slow for those engaged on the work. At long last they managed, after prodigious efforts, to open up a path, only to find the gold as far off as ever. It was buried many feet deep in sand and mud, and to dig it out with shovels was an impossibility, for the sea would wash the sand in just as quickly as the divers shovelled it out.

Forty yards above them lay the *Racer*—a floating workshop full of the most remarkable inventions that scientists and engineers could devise to assist submarine work. Aboard was a mighty 18-inch pump capable of sucking up a mountain of sand an hour. The mouth of this monster appeared from above. It was placed in position by the divers, and they watched the silt melting before it as if by magic, flowing up to the surface to be dumped a little distance away.

It is no uncommon thing to find such a pump sucking up chunks of rock weighing half a hundred-weight, and even trying to remove bits of girder and plate. But such objects, like deck planks, are rather apt to stick in the bend, and then the monster chokes and has to receive the attentions of the salvors.

Remarkable as was the work done by the gallant divers, the results of the season's work were fearfully disappointing, for only seven bars of gold worth about £10,000 in all were recovered. In no wise discouraged, the treasure-hunters stole back to the old spot the following spring to try their luck again. The gales of the winter had torn great plates from the wreckage as though they were merely sheets of brown paper and dropped them yards away; the decks that had once resounded to the laughter of beautiful women were laid down flat with the seabed. Twisted and rusted iron lay for hundred of yards around. Looking for a needle in a haystack were an easy task compared with finding the treasure amid all this tangled debris.

A long, keen search revealed what had once been the

strong-room. Great metal plates were piled over it, necessitating blasting operations once more. The divers toiled until the plates were cut and dragged away. Then incredible quantities of silt had to be eaten away by the sand-pump, the divers watching closely and coming on a bar from time to time. By the end of August, 1922, gold worth £150,000 had been secured, and early one morning H.M.S. *Wrestler* might have been seen slipping into Liverpool. Directly she moored beside the quay, case after case was landed from her and placed in a motor-lorry. Those cases—a dozen in all—were full of gold which had been recovered from the *Laurentic*, and each case represented a small fortune.

All through the season of 1923 the divers carried on, searching amid that chaos of rusted iron for the gold and silver bars, wresting them one by one from their hiding-places on the seabed. For seven seasons they have fought the ocean for that mighty fortune of over £5,000,000 and their heroic efforts have led to the recovery of £4,750,000. Considering the depth in which the *Laurentic* sank, and the perils and difficulties besetting the workers, the results are beyond compare.

Never before has there been a treasure-hunt of such magnitude, and how long this will last no one can say. A big fortune of £250,000 still lies hidden just off the coast of the Irish Free State, and, if the British Navy fails to recover it for the British Treasury, it will be for the simple reason that its recovery is humanly impossible.

For every £100 won back from the depths, the divers have received an award of 2s. 6d., so altogether they have shared among themselves the sum of £5,937 a sum that has been well and truly earned. It says much for the efficiency of the British Navy when it is known that the whole of this perilous treasure-hunt has been carried out without a single accident to any of the divers engaged.

Many rumours have arisen of wonderful machines being used to locate the treasure, of instruments with the power to divine the presence of gold, of scientists standing on the deck of the salvage vessel watching, with bated breath, a needle oscillate round a dial until it has indicated that the diver far below is in the vicinity of the precious metal. These rumours, however, have no foundation in fact, for the treasure has been recovered solely by straightforward diving. The estimates of the treasure sunk have also varied from £3,000,000 to £8,000,000, but the figures given here have been furnished me specially by the Admiralty, and they are therefore strictly accurate.

# CHAPTER VII

British salvage experts have performed extraordinary feats; the American Navy has produced divers excelling even our own; but it has been left to the Italians to accomplish the seemingly impossible. As a sheer feat of salvage, the raising of the *Leonardo da Vinci* remains unsurpassed.

The night of August 2, 1916, will long be remembered in Taranto, for just before midnight the whole town was awakened by a tremendous explosion. The people leapt from their beds and rushed towards the harbour, to find searchlights sweeping the bay and the finest battleship in the Italian Navy belching forth flames and smoke. The *Leonardo da Vinci* was doomed. In a moment 250 officers and men were wiped out of existence, and although the survivors fought most valiantly to quell the fire that enveloped the ship their efforts were vain.

Suddenly the decks of the battleship canted beneath them, shooting them like flies into the bay, and she swung right over and sank upside-down in 36 feet of water. The searchlights from the surrounding battleships lit up the darkness. Round and round they flashed, seeking the enemy who had dealt this mortal blow; but there was no sign of a periscope, nothing but the heads of the Italian sailors fighting for their lives in the sea.

A time bomb, secretly introduced into one of the magazines, had robbed the Allies of one of their most powerful battleships. This loss of a first-class ship of 24,000 tons, equipped with an armament of thirteen 12-inch guns, was a grave one to the Italian Navy, and the question of salving her at once arose. Famous foreign experts came on the scene, gazed on the visible portion of the keel of the ship which had cost £4,000,000, and shook their heads dubiously.

"Impossible!" they said. "The only thing to do is to blow her to pieces."

The eyes of the Italians flashed. Somehow, at some time, they determined to salve the battleship. It might be impossible during the war, owing to the difficulty of getting material for the operations, but in their own minds the honour of Italy would never be satisfied until the ship which lay at the bottom of Taranto bay once more floated on the seas.

The sinking of the *Leonardo da Vinci* was, indeed, a great blow to the pride of the Italian Navy, and there was a general desire on the part of the nation to wipe out the stain and turn defeat into a triumph by refloating the ship. The more difficult the task, the greater the triumph; the more impossible it seemed to foreign experts, the more determined were the Italians to achieve it.

Throwing themselves heart and soul into the matter, the officers of the Italian Naval Engineering Corps studied the problem most carefully and formulated several schemes, among them a plan to build around the ship a floating dock which, when completely pumped out, would automatically lift the wreck. Shortage of steel and other materials at that time made this plan impracticable. Then General Ferrati, the chief of the Italian naval constructors, evolved a plan to raise the ship by means of compressed air and carry her upside-down to the dry dock at Taranto, where she could be prepared for righting.

It must never be forgotten that the battleship was upside-down, and that not only had she to be raised, but she also had to be righted. Rivet by rivet and plate by plate she had in the course of years been built up by hundreds of men into one of the strongest structures known. All the rivets and plates had been welded into a compact mass of 24,000 tons which now lay at the bottom of the sea. Afloat, she obeyed the hand and brain of man, would go wherever he desired; at his behest she turned to right or left, sped furiously through the sea or stopped. Now she was immovable as the mountains; to smash her to pieces would have been a gigantic task, costing months of time, tons of much-wanted explosives, and well over £100,000 in money. The queer thing is that Ferrati proposed to harness air to lift the sunken monster, just as though she were an airship instead of a battleship. In such ways do master-minds work.

So brilliantly conceived were Ferrati's plans that orders were at once given to put them into execution. Divers went down to make a survey of the wreck, which was so rent by the explosion that a vast hole had been blown right through her from keel to top deck. A further survey indicated that the huge ship was literally digging her own grave. The weight of the upside-down battleship was all resting on the funnels and gun turrets, and these, owing to the enormous pressure from above, were piercing a way slowly but surely through the mud. Day by day the ship sank lower and lower, until the whole of her upper deck was completely buried and the greater part of her hull at the stern had disappeared. In six months the funnels cut down through a bed of mud over 30 feet thick before they encountered a bed of clay, which arrested the sinking of the ship.

The Italians bringing the *Leonardo da Vinci* upside down into dock at Taranto on September 17, 1919, after fighting for over two years to raise her from the seabed.

No wonder the experts gave up hope. It really seemed that nothing but a miracle could bring the great vessel to the surface again. There she was, upside-down, buried deep in the clinging mud, an enormous, unwieldly mass that the biggest cranes ever invented were powerless to lift. It is a comparatively easy task to raise a weight of 10 tons from the seabed, but it is quite a different proposition to lift a mountain of metal weighing upwards of 20,000 tons.

The upside-down battleship safely docked on September, 18, 1919, with the giant pontoons which helped to raise her still lashed to her sides.

In no wise discouraged by the difficulties of the problem, General Ferrati and his associate, Major Gianelli, ordered large-sized models of the ship to be built. These were accurately constructed down to the smallest detail, with miniature engines, propellers and guns; and every compartment was loaded to represent the things on board the battleship when she foundered.

A stranger might have laughed at the childishness of the Italian officers who were apparently playing with toy battleships. But things are not always what they seem. Actually these same officers were puzzling out the most abstruse problems, carrying out remarkable experiments which enabled them to determine how the ship should behave in certain circumstances. As a result were evolved some intricate calculations upon which depended the whole operation of raising the ship.

The small part of the keel still showing above the surface was used as a platform on which to build huts for the salvage workers. Other huts were erected, in due course, on platforms built up from the submerged keel. The assembling of the plant for the work was completed by the spring of 1917, when the people of Taranto began to observe the figures of divers about the wreck.

Those divers had no enviable time. They quickly discovered that the explosion had liberated a quantity of thick oil which clung to everything within the ship, and as they went down it obscured the glass of their helmets and rendered the men practically blind. As if the oil were not sufficient handicap, there were thick clouds of rust which fogged the water and added to the discomfort of the divers. Yet the oil, despite its drawbacks, proved something of a blessing, for it adhered to hundreds of shells and protected them so efficiently from the action of the sea that the Italians were able to use them after salving them!

The recovery of the ammunition was the first step to lightening the ship. Day after day shells were hoisted out of the wreck and loaded into lighters. It was dangerous work, but it became rather monotonous to those engaged in it. Monotony, as is well known, is apt to lead to carelessness, and carelessness in handling shells may lead to terrible results. It is a fine tribute to the carefulness of the men engaged on the work to know that they salved nearly a thousand 12-inch shells, three thousand 4·7-inch shells, some torpedoes, thousands of explosive charges and hundreds of tons of other ammunition without a single mishap.

Meanwhile, a cable was laid from the power station at Taranto right out to the wreck, a distance of a mile and a half; and with the power thus furnished the divers began drilling holes to take the rivets that were to hold the patches over the great rents in the hull. Slow and arduous work it was, and not without danger, for it cost one man his life. The patches were lowered into place, a layer of rubber was fitted betwixt the hull and the edges of the patches to make them watertight, then the patches were successfully bolted home.

More cables were carried out from the power station to work the air compressors, and, as soon as the divers had made a number of compartments watertight, the salvors began to pump air into the sunken vessel. The air which was pumped in naturally rose. It tried to get away to the surface, but the keel of the battleship, which had been most carefully repaired and made airtight, prevented it from escaping.

The air was thus caught, as it were, in a trap. There was no way out for it. It was not strong enough to break through the bottom of the ship, but it was strong enough to press down the water within. As the volume of air increased, the belt which it formed grew in depth until it had forced the water down for a distance of 26 feet below the level of the sea outside, and men were able to enter the bottom of the vessel through an air-lock, work in security in this belt of compressed air, and lighten the vessel by taking out her stores and coals.

By the beginning of November, 1917, the salvors occasionally felt the battleship stir slightly beneath their feet. Despite the fact that she was buried deeply in the mud, her bow was showing the slightest of inclinations to rise. The engineer in charge noted this with delight. Barely perceptible as was the movement, it was more than sufficient to encourage him to persevere.

Once more the thick oil cropped up to hamper operations and increase the many difficulties. As the water was forced down inside the vessel by the compressed air, the oil was deposited on everything. In most cases this did not matter much, but it was of far-reaching importance when it came to searching for leaks in the hull. The oil so obscured these places that it was extremely difficult to locate them, yet everything depended on their being discovered, for had they been left unstopped they might have let out the air and made it impossible to refloat the ship, or, alternatively, let in the water at a critical time and led to her sinking in such a position that she could never be floated again. Fortunately, the Italian salvage men were able to detect all the leaks and stop them effectively, as the sequel amply proved.

Critics of the operations pointed out that, should the salvors succeed in floating the battleship upside-down, there was not sufficient depth of water to allow her to be taken across that mile and a half of sea to dry dock. Even if they managed to get her to dry dock, all their work would be wasted, for the battleship floating upside-down would draw at least 50 feet of water, and the dry dock at Taranto was only 40 feet deep.

These difficulties were fully considered and plans made for overcoming them. As it was an impossibility to increase the depth of the dry dock, the only way to solve this problem was to decrease the depth of water that the battleship would draw. The engineer accordingly proposed to detach the funnels, gun-turrets and other top hamper from the deck of the vessel.

After floating for two days in dock, the battleship was coaxed into position until she settled without accident on the wonderful timber framework shown here. It was a fine feat to accomplish.

A remarkable photograph taken from an airship of the upside-down battleship in dry dock.

So firmly embedded were these things in the mud, that the feat of cutting them off appeared to be more than mortal man could accomplish. It was, too, pointed out that if divers tried to clear the mud away from round the funnels, to enable them to work at their task, the sea would quickly fill up the cavities again. Yet another aspect of the problem was that the mud pressing upwards against the deck of the battleship was preventing her from sinking deeper, and if the mud were removed the whole weight of the *Leonardo da Vinci* would once more rest on her funnels and turrets and drive them deeper still into the clay.

But the engineer, with a stroke of genius, made no attempt to clear away the mud at all. Instead, he tackled the job from inside the ship. Certain compartments were pumped out and used as air-locks, and in one turret the salvors succeeded, by the use of compressed air, in lowering the water to a level of 56 feet below the surface of the sea.

The men who performed the mighty task of detaching the turrets from the ship actually worked 20 feet below the level of the mud. All around them outside was 20 feet of thick black ooze, and above that the illimitable ocean; yet the air we breathe, properly compressed, held back the deadly waters and enabled the men to work in safety. No wonder the experts say we are only just beginning to discover the remarkable power of compressed air as an aid to salving ships!

Throughout 1918, some 150 men laboured about the ship to free her from her top hamper and masts. Despite all difficulties, the gun-turrets, funnels and other deck projections were detached from the ship and specially prepared so that when the vessel was raised they, too, could be brought to the surface. The open spaces in the deck left by funnels and turrets were covered in and made quite watertight, scores of tons of cork being packed into the *Leonardo da Vinci* to give her buoyancy.

Early in 1919 one or two tests showed that they could raise the monster when the time was ripe. But Major Gianelli, the engineer in charge, was taking no chances. To make quite sure of lifting her, he caused eight large pontoons to be fixed to her, each capable of sustaining a load of 350 tons, so in all he obtained from them the power to lift 2800 tons. These pontoons, or camels as they are sometimes called in salvage circles, are strong metal cylinders something like big boilers or tanks. They are of the utmost importance in salvage operations and figure in most wreck-raising work. All were filled with water and sunk into position exactly where their lifting power was most wanted. The divers lashed them with strong steel cables securely to the sides of the battleship, and by the month of June the work on the mammoth craft was all but complete.

Remained the problem of making it possible to tow her to dry dock. Notwithstanding that all projections had been cut away from her deck, she drew so great a depth of water that it was obvious she would foul the bottom before going any distance. To obviate this danger, the Italians set dredgers to work to cut a channel all the way from the wreck to the gates of the dry dock. The making of this channel, which was a mile and a half long, entailed the removal of thousands of tons of mud, but the salvors regarded this task as trivial compared with the work they had accomplished on the overturned ship.

Then the dock itself required to be specially prepared, for like all dry docks it was planned to take a vessel upright and not upside-down. The chocks down the centre of the dock, which normally support the keel of a docked vessel, were quite useless so far as the *Leonardo da Vinci* was concerned. So a forest of timber began to spring up in the dry dock. Mighty baulks of wood, 15 inches and more square, were built up from the bottom of the dock. These followed the outline of the ship so that the deck could be brought exactly over them and allowed to sink into place upon them. Other gigantic piles of timber were constructed to support particular parts of the deck.

By September 17, 1919, all these preparations were completed. The air compressors forced the water out of the pontoons and out of the hull. Certain compartments of the ship were filled with water in order to balance her evenly—and then the keel, with the great pontoons straining upwards, slowly arose out of the sea. For a time a stern battle went on between the mud which was gripping her and seeking to hold her down and the air which was striving to lift her to the surface. Then the air won. The battleship slipped from the grip of the mud, leaving her guns and turrets still embedded, and floated on the surface once more.

A unique photograph of the *Leonardo da Vinci* as she lay in the Bay of Taranto with all the salvage craft around her just before she was turned over.

A rapid survey was made to see that she was fit for her journey, then the tugs took up their task and began to tow her slowly along the channel between the lines of buoys marking the passage. A stranger spectacle than the towing of this upside-down battleship was never before seen on the seas. The tugs managed to keep the capsized leviathan right in the centre of the channel, and by nightfall the vessel was at the entrance to the dry dock, and was skilfully manœuvred inside on the following day.

For two days she floated, held up by the compressed air within her hull, and during this time certain adjustments were made in the mighty timber frame that was to support her. The water was now drawn off from the dock and the *Leonardo da Vinci* settled down comfortably on her timber framework.

Her settling down placed a huge strain on the timbers, some having to bear the very great pressure of 225 tons to the square inch. The calculations, however, were so cleverly made, and the vast weight was so evenly distributed, that the framework supported her in perfect security. In itself this was a remarkable achievement. The slightest miscalculation, or one weak timber, might have brought about the collapse of the whole structure, and the battleship would have fallen, an absolute wreck, on the bed of the dry dock.

Towing the upside-down battleship out of dock on January 22, 1921 in order to right her.

For months men swarmed about the upturned battleship, doing the final repairs that were necessary before she could be righted. The conclusive test of the Italians was nigh. Could they succeed in turning the great mass of metal the right way up again? No power known to man would suffice to right the vessel on land. Before the task could be attempted it was essential to place her once more in her element, the sea. On land she was immovable, on the sea she floated and could be more or less controlled by man, but whether man could perform the miracle of turning her right way up again, nobody knew.

The bottom of a ship, of course, has to be strongly built to withstand the pressures to which it is subjected. The deck, not having to stand the strain that the bottom is called upon to bear, need not be built so strongly. In this case the deck and the bottom had changed places, and it was therefore of the utmost importance that the deck should be strengthened to withstand the increased pressures that would arise in righting the ship.

Out in the bay the dredgers scooped a deep basin to enable her to turn over without fouling the seabed, and towards the end of January, 1921, the *Leonardo da Vinci* was towed to the place where it was proposed to right her. Four hundred tons of solid ballast had been loaded into her, and the engineers made preparations for pumping 7500 tons of water into certain compartments on her starboard side. Being above the centre of gravity, this weight would make her so top-heavy that she was bound to overbalance and thus turn right side up again.

Upright once more after being upside down for four years. she raised a huge wave as she swung over, as may be seen from this photograph which was taken from an airship

There in the bay lay the still stricken leviathan. The valves were opened to allow the sea to enter her compartments, and the salvage men scrambled from the upturned keel and pulled away from her in their boats. The water began to flow in, and by the time some 800 tons had entered she began to turn ever so slowly. Soon, as the weight of water increased, she swung over with a rush, raising a big wave as the deck swept clear of the water. For a moment it looked as though she would swing right over and finish upside-down again. But the engineers had worked out their calculations to such a nicety that the battleship finally came to rest with a slight list, just as they had foreseen.

Across her deck, in big letters, was seen the motto of the famous *Leonardo da Vinci*: "Every wrong rights itself," painted while the vessel was still upside-down in dry dock. It was a happy thought, and a pandemonium of cheering broke out as the legend came into view to tell of the most remarkable salvage feat ever accomplished.

The *Leonardo da Vinci* ready to go into dry dock again to be refitted. A brilliant salvage feat is recorded in these remarkable photographs which are reproduced by courtesy of the Italian naval attaché.

The salving of the ship and her final righting took four and a half years. It was a Herculean task, and from first to last cost the Italian Government £135,000. Unhappily, General Ferrati, who conceived the brilliant plan, did not live to see it completed. He was succeeded as director of operations by General Faruffini, who in turn was succeeded by General Carpi, but during the whole time Major Gianelli was in charge of the work and to him is due the credit for carrying out from beginning to end, and bringing to a triumphant conclusion, the most wonderful salvage feat ever performed by man.

# CHAPTER VIII

Before the Great War the number of concerns specializing in salvage work were so few that probably all could be numbered on the fingers of both hands. Sweden had a fine salvage unit at Stockholm, a Danish company worked from Copenhagen, Germany possessed a very powerful salvage plant, while perhaps half a dozen salvage concerns operated in British waters, the most important being the Liverpool Salvage Association, the London Salvage Association and the famous firm of Henry Ensor, at Queenstown in Ireland.

In the number of marine salvage units she could muster, Great Britain was thus particularly fortunate. The dangers of our coasts have long been regarded as a drawback, yet in time of crisis they proved a blessing in disguise, for the yearly toll of wrecks on our shores has provided fine experience for our salvage experts and made them second to none in the world.

When the Germans hurled their challenge at humanity, all the salvage concerns operating in Great Britain were at once taken over by the Admiralty and placed under the command of Commodore F. W. Young. For long Commodore Young had acted as chief salvage officer to the Liverpool Salvage Association, and forty years' experience of raising wrecks had given him a knowledge of the subject that was unique. Wandering round our shores in storm and shine, fighting to get ships off the rocks, struggling to save their cargoes, he learned to know our rugged coast better than the average man knows the lines on the palm of his hand. The reefs from which a ship might never escape, the sandy bays that provided shelter, the bars that lurked in wait for unwary ships, all were known to him. His knowledge was such that he was able to sum up the chances of a ship directly he heard where she was wrecked.

Whatever blunders may have been made in appointing

other men to other commands, the First Lords of the Admiralty made no mistake in selecting Commodore Young to be Director of Naval Salvage. Generals came and went, Admirals were moved up and down, but this one man was in control of the Admiralty Salvage Section throughout the whole war, bearing the grave responsibilities of a most important post from beginning to end.

The first work of the Admiralty Salvage Section was purely naval. These were the men who laid the mines to guard our harbours, and upon them devolved the duty of laying down those long lanes of mighty nets to protect our troopships hurrying from England to France. When the *Lion* was so sorely stricken at Jutland, it was one of the section's salvage steamers that helped her into port, and they were men of the Salvage Section who patched her scars and made it possible for her to limp home.

But the work of the Salvage Section changed completely with the coming of the unrestricted campaign of the German submarines. No longer was it purely naval in character. Thenceforward it became general, and the officers and men of the section had to stand ready to save merchant vessels as well as warships.

So grave a menace was the enemy submarine campaign that foreign shipowners refused to take the risks of sending ships to Great Britain, for no underwriter with any sense could be expected to insure ships when the Germans were torpedoing merchantmen at sight. Similarly no shipowner with any sense would send a ship here that was uninsured, for if his ship were torpedoed the whole loss would fall on him. For this reason alone there was a likelihood of diminishing supplies of food and munitions coming to our ports.

The British Government rose to the situation by becoming the biggest underwriting concern in the world and insuring every ship entering and leaving our ports. Great Britain accepted the responsibility for all losses, and the shipowners knew they were sure to get their money in the event of their ships being sunk. As a further precaution, the system of convoy was instituted, whereby half a dozen or a dozen ships journeyed together under the escort of some of our warships. An additional measure to cope with the marauding submarines was to arm our merchantmen so that they stood at least a chance of beating off an attack.

Shrewd as were the German calculations of winning the war by the submarine campaign, and nearly as the enemy succeeded, they reckoned without our Admiralty Salvage Section. While all the powers of the British Admiralty were concentrated on destroying the German underwater craft, the abilities of the Naval Salvage Section were focused on repairing the damage wrought by enemy torpedoes. From a comparatively minor position, the Salvage Section sprang into paramount importance. As the list of torpedoed vessels grew day by day, so our salvage organization was enlarged to grapple with the extra duties.

Directly a ship was torpedoed, the news was wirelessed to Whitehall, and the nearest available naval craft was ordered to stand by and render all the assistance possible until a salvage steamer arrived from the most convenient depot to take over. Salvage steamers and depots were dotted at various ports all round the coast, and as soon as particulars flashed through to the Director of Salvage he detailed his nearest available unit for the job. If a vessel still floated, he despatched powerful tugs to tow her to port; if she sank, he instructed a salvage officer to report on her position immediately.

No time was wasted, for the loss of one tide might easily have meant the total loss of the vessel. Within a few minutes of the report coming to hand, the Director dealt with the case and suggested how it should be treated.

Commodore Sir Frederick Young's calmness was indeed amazing. I have vivid recollections of seeing him in his room at Whitehall when the submarine campaign was at its height. The newspapers were full of the tales of sinking ships, people were talking about it agitatedly, faces in the inner precincts of Whitehall were grave and obviously concerned, but the Director of Salvage remained quite unruffled. As I sat talking with him, the news came through of seven more ships being sunk; on top of it arrived the information that one of the salvage ships herself had been torpedoed in the Mediterranean. Yet the Director of Salvage did not turn a hair.

He asked one of his officers the whereabouts of another

salvage craft.

The officer told him.

"Send her out to replace the ——," and he mentioned the name of the sunken salvage ship, which I have long since forgotten.

He puffed quietly at his pipe, screwed a monocle into his eye, and scanned the telegrams with their bad news. Then he gave a few orders, and in a moment or two the wires were humming with instructions to various salvage units to hurry to the aid of the stricken ships.

It was all done so quietly and simply, without one sign of flurry or fuss on the part of the sturdy figure clad in a simple blue serge suit such as thousands of civilians wear to-day. Yet coming in and out and waiting deferentially on his word were naval figures resplendent in gold braid. The contrast emphasized the simplicity of the man controlling this supreme service. His unaffected ways and quiet manner masked an amazing cleverness, for no man alive was imbued with a greater genius for sea salvage work than this modest man sitting composedly at his desk by the pleasant window in Whitehall.

His big room was set off in the centre by a round polished table containing a bowl of flowers. Photographs of salvage ships dotted the walls, while various charts of the British Isles stuck full of coloured flags bristled with information to those able to read them. Other charts were concealed beneath spring blinds that sprang up at the touch of authority. By studying these charts, the Commodore was able to tell at a glance just how the situation stood, where ships were sunk, where ships were beached, where his salvage units were working. On a side-table was a big book of charts that could only be lifted with an effort, and another table contained a model ship showing the standard patch.

Called into being by the war, the standard patch certainly proved one of the greatest aids of the Salvage Section, for many a ship that would have ended her days at the bottom of the sea was brought safely into port under the protection afforded by this invention. The standard patch was formed of grooved timbers fitting one into another, something like matchboards, and in appearance it resembled the top of a gigantic roll-top desk. Owing to its construction, it was admirably adapted for fitting the curves of the hull of a ship.

A torpedoed ship which was saved by being beached.

In fitting a standard patch, the size of the hole in the hull was first ascertained, then the patch was made, bolted into position, and the edges were made watertight with cement. Many ships had to be beached at the nearest spot in order to save them from foundering, and the standard patch was then fitted to enable them to reach port and undergo permanent repairs. Other ships still remained afloat after being torpedoed, and it was no uncommon sight to see the ship's carpenters constructing a standard patch upon the deck. When the patch was finished, it was lowered over the side, the bottom edge being weighted to make it sink in an upright position, while the divers guided it into place and secured it with bolts and nuts.

Despite its temporary character, the repair was strong enough to enable the ship to journey to the dock set aside for her reception. Yet many a ship met various adventures on the way, and her journey to port was rather a protracted affair. One such case was that of a large vessel torpedoed by the Germans. Luckily, she did not sink immediately. Her bulkheads held and her captain was able to head for the shore until she touched bottom and settled down. Along came the salvage unit, and, ascertaining the damage, worked desperately to fit a standard patch. The patch was duly put on, the many bolts screwed up, and the vessel pumped out and towed off to port.

The standard patch which was fitted over the hole in the ship's side. As may be seen, each timber was bolted home and the edges were made watertight with cement. these patches were of great service in defeating the German submarine campaign.

The salvage officers were congratulating themselves on work well done when the unexpected happened. There was a dull explosion and a giant cascade against the side of the steamer. She had been caught a second time by a German submarine! Her nose was headed inshore and once more she touched bottom.

Quickly as they could, the salvors tackled her, for she was not the only ship on the sea receiving the unwelcome attentions of the Germans, and the salvors were in constant demand all along the coast. They sized up the new damage, made another patch, drilled the holes in the hull, fitted a felt bed for the patch to rest against and screwed it tightly home. Then the pumps were set going, the damaged hold was emptied and her keel came up from the sandy bed in which it had been resting.

The ship, which had survived two German torpedoes, continued her interrupted journey, but she had only been an hour or two on the way when another enemy submarine got her. Whatever the salvage men said and thought, they started to patch her up again, and in time they had the thrice torpedoed vessel continuing her slow journey to the dock where she was to be repaired.

Before they could get her home, however, her rescuers were compelled to beach her and struggle to save one or two very urgent cases. They accordingly put her ashore in a sheltered bay in the Isle of Wight where they knew she would be quite safe until such time as they could attend to her. She was months making a short trip of a few miles round the south coast, but she seemed to have as many lives as a cat, and eventually reached dry dock where the damage wrought by German torpedoes was properly repaired.

The remarks of the Germans must have been rather interesting when they discovered that they were torpedoing the same ship time after time. Probably they thought it was some trick the British were playing on them, some gigantic bluff to make them waste torpedoes. Anyway, although they tried and tried and tried again, the Admiralty salvage men, not to be outdone, managed to save the ship from the clutches of the Germans after all.

So long as the submarine campaign continued, it was indeed a gigantic tussle between pumps and patches and torpedoes. At first the torpedoes had it all their own way, but pumps and patches in the skilful hands of the Admiralty Salvage Section began to rob the Germans of more and more of their prizes, and they ultimately proved a most important factor in bringing home to the foe that the game was not worth the candle.

The demand for pumps of all types was tremendous. Motor pumps, steam pumps, electric pumps—all were required, and the pump-makers were kept busily employed night and day. The war brought out the good points of one pump known as the electric submersible pump. Invented in pre-war days by an electrical engineer named Macdonald, this invention did not attract the notice it deserved, and in the end the inventor sold out his rights and emigrated to Canada. Since then his pump must have been very successful financially, for one or two that happened to be aboard a battleship at the battle of Jutland did such wonderful service that the whole of the British Navy was fitted with them.

Many had tried to solve the problem of an electric pump, but generally they came to grief owing to the current short-circuiting in the water. Macdonald worked at the problem until he succeeded in overcoming it, and the result was a drum-like pump with the inner parts spinning at a high speed and forcing the water upwards through the pipe. Instead of fixing his pump at the top end of a suction pipe, Macdonald placed his pump at the bottom end of a pipe and dropped it into the water. The pump weighed about half a ton, and owing to the fact that it worked entirely under water, with water flowing all round and through its bearings, it was not liable to suffer loss of efficiency through air leakage. The tendency of the pump to overheat owing to the speed at which it worked was checked by the cold sea water always passing through it. It was, in effect, a water-cooled pump that was excellent for working at depths a little beyond the reach of the ordinary pump.

For touch-and-go cases the submersible pump was much in demand by salvage officers, but for cases that required long and steady pumping for days and perhaps weeks the wonderful Gwynne pumps were not to be excelled. Their extraordinary reliability is marvellous. So long as you give them the steam to work with, coupled with proper attention, they will do almost anything that you ask of them. They will pump steadily for days and even weeks without stopping, throwing overboard the specified number of tons of water an hour. They are, indeed, among the mechanical marvels of the age, practically as perfect as any machine is ever likely to be.

So sure are they, that salvage men will willingly put to sea in a badly leaking ship and set out on a voyage that may last a week or two. If the pumps stopped, the ship might founder in two or three hours. The men know it, but they do not worry. They have implicit faith in the pump, and although merely the power of the pump stands between them and death they carry on quite unconcerned. And while the water is finding its way into the breaches in the hull of their ship the pumps are steadily throwing it over the side.

As Henry Ensor, one of the cleverest salvage experts

alive, once remarked to me: "For a long voyage in a leaking ship, give me the Gwynne."

Three of the electric pumps which proved their efficiency during the war. They remained in the hold of the sunken *Westmoreland* for three months until she was raised. Worth £3,000,000, she was by far the richest salvage prize of the whole war.

Pumps, indeed, played a big part in beating the German submarine, and it was the submersible type that figured in the case of the *Westmoreland*, for three placed in the hold of this vessel were left submerged for nearly three months and upon withdrawal worked quite as well as when they were put down.

The damage wrought by a German torpedo. A good idea of the immensity of the hole may be gained by comparing it with the legs of the man standing on the scaffolding in the wrecked engine room. Despite the damage, the ship was saved.

No richer prize than the *Westmoreland* fell to the Salvage Section during the whole war, for ship and cargo were worth about £3,000,000. The vessel was steaming in the neighbourhood of St. Bees Head on her way to Liverpool when an enemy submarine let loose a torpedo. The missile ran true, and a moment later a terrific explosion told the Germans they had bagged their game. Whereat the attacking submarine, knowing the sea thereabouts was likely to be well patrolled for some little time to come, quietly slid off.

True as the torpedo ran, the Germans made a slight miscalculation. Though trifling, it made all the difference in the end. Instead of the torpedo hitting in that vital spot amidships and destroying the engines, it struck forward in No. 2 hold and tore an enormous hole in the hull of the ship big enough to drop a small house into. The heart of the ship, the engine-room, was untouched, and the captain still retained the power to drive his ship through the seas.

Slim destroyers slipped over the horizon and crowded round the torpedoed vessel. Fortunately her bulkheads held firm and, although the damage was such that it looked as if the ship might founder at any moment, the captain held his course in a valiant attempt to reach port.

Slowly the bow of the ship sank lower and lower in the water, until it seemed impossible for her longer to remain afloat. At last a destroyer manœuvred into position and took off captain and crew, and they stood by to see the last of the ship. Instead of sinking, however, she still hung there, and the captain and crew returned to her in order to try once more to get her to port. There was just a chance that they might succeed, and the captain was not going to lose that chance.

Engineers and stokers went below to give her steam, and she limped lamely along, continuing to go down by the head. As her bow went down, so her stern came up until it was obvious that if she did not soon sink she was bound to become unmanageable, for in a short time her screws would be clear of the water and churning the air instead of the sea. Heading her for the beach while there was yet time, the captain took her in until her propellers were right in the air and her bow scraped the bottom, then he and the crew were taken off and the *Westmoreland* quietly settled down.

If only she had settled at high tide, the *Westmoreland* might have proved an easy case for the Salvage Section to deal with. But with the usual perversity of things, she went down at low water, and as the tide rose, the sea began to pour out of the broken hold along the shelter deck and over the tops of the bulkheads into all the other holds. Unluckily her bulkheads had not been built right up to the top deck. Instead, they reached only to the previous deck, the shelter deck, and there was nothing to prevent the seas washing the whole length of the shelter deck, which was just what they did. The consequence was that the whole ship filled with water, and at high tide she was quite submerged, with her top deck 30 feet below the surface.

Commander Kay hastened to the spot and surveyed the wreck. Quickly he saw that the only way of raising the ship and getting her to port was to prevent the seas from washing out of the damaged hold into the sound holds. It appeared simple, but the men who began to strive to carry out the scheme had the struggle of their lives.

It was February, when the weather was just as bad as it could be. The heavy seas and strong currents effectually prevented any work being done for three or four days a week, and on the other days it was only possible to work for two or three hours at low tide. Watching their opportunities, the divers scrambled into the wreck and gradually timbered in a mighty hole, 40 feet across, that was blown in the shelter deck by the force of the explosion. The first step in their struggle with the sea was looked upon as won.

Barely was the work completed when the sea, frothing with fury, raged through the hole in the hull and battered continuously at the underside of the work until the timbering was reduced to matchwood. I have already mentioned that salvage men are sparing of words, and, if they said but little on this occasion, no doubt what they said was to the point.

With that patience which is beyond all praise, they resumed their efforts with a firm determination not to be again cheated by the sea, so they used steel to counter the force of the waves. Whenever tide and weather served, they worked with might and main to build watertight walls—or a steel trunkway, as the salvors called it—from the shelter deck of the damaged hold right up to the top deck in order to confine the sea to that hold and prevent it from washing over the tops of the other bulkheads. By then the salvors realized that it was quite hopeless to attempt to patch the hull of the ship to prevent the seas from entering, for no temporary work could withstand the full force of the Atlantic gales. Consequently, the divers concentrated on building their trunkway, and in a month it was completed and the water was effectually shut off from washing into the other holds.

The salvors determined now to try to move the ship to a more sheltered position where they would be able to work for longer periods and with fewer interruptions. Accordingly, pumps were set to work pumping out the water in the sound holds, and in time the *Westmoreland* swung clear of the bottom. The tugs caught hold of her and towed her inshore for a couple of miles, when she bumped the bottom again and was allowed to settle. It was 2 miles to the good, the water was much shallower, but even more important was the additional shelter which made it possible for the men to work more continuously.

So the divers toiled away with renewed vigour, hauling the cargo out of the ship to lighten her, hoisting out case after case of butter for which the people were clamouring. It was, fortunately, none the worse for its immersion, and I believe it duly reached the tables of the people, who had no idea that they were eating butter which had been at the bottom of the sea. If the true story be told, there is little doubt that a large quantity of food rescued from the clutches of Neptune was duly eaten by the British people without their being any the wiser. Necessity knows no law, and when famine is looming nigh, as it was then, butter that has been on the seabed is better than butterless bread. In any case the butter ration was so small—but two ounces a week—that no danger could possibly accrue through eating it.

Tons and tons of timber props were built into the ship to strengthen her in all directions. The problem of patching the vessel was again considered, but the weather was such as to render patching impracticable. So the salvors allowed the waves to thunder in through the gaping hole in her side, whence they gushed out of the top of the ship in fountains of spray. There was nothing else to be done in the circumstance. Had the salvors succeeded in covering in that mighty hole in the shelter deck strongly enough to keep back the seas, the seas would have raged about inside the damaged hold and smashed everything to pieces; consequently it was much wiser to leave them an outlet. The trunkway was a safety valve by which the seas escaped after tearing through the gaping wound.

Once the forward holds of a ship fill and drag her down by the bow she is rendered helpless. She may still remain afloat, her engines may be perfect, but her captain no longer has any control over her because her propeller is out of water.

Fourteen weeks after work was first started, Commander Kay decided that the time had come to make the final lift and get the *Westmoreland* to dry dock. The electric pumps were switched on and kept running until the waterlogged holds were cleared, and the torpedoed vessel rose off the sandy bottom and floated. Then cropped up the vital matter of balance. For weeks the divers had been fighting to rid the ship of water, and now, paradoxically enough, they found they had pumped out so much that her stern came up clear of the surface, while her bow was barely clear of the sand.

It was useless to attempt to tow her to port under such conditions, for in a short while she would have been digging her nose into the sand and sinking once more. Before the journey could be essayed, it was essential to balance her properly, and this could only be done by leaving a sufficient weight of water in the after holds to balance the water in the forward hold. They had to trim the ship by using water as ballast. Calmly they allowed the after holds to fill again, then they set the pumps going until she rose on an even keel. The stumpy tugs fastened on to her and did not let her go until she was safely in dock.

Altogether the Admiralty Salvage Section during the war salved nearly 500 ships, valued with their cargoes at about £50,000,000. While the submarine campaign continued, the British need for shipping was so great that all salvage efforts were concentrated on those ships that could be quickly salved and put into commission again. The easiest cases were dealt with first, and the more difficult cases were left until there was a reasonable opportunity of coping with them.

A careful list compiled by the Admiralty after the war showed that there were 416 war wrecks lying in less than 20 fathoms, or 120 feet, around the British coast, and of these it was estimated that one in ten might perhaps be raised. Actually 51 war wrecks were salved after the Armistice, but as some of these were lost in foreign parts the original estimate was not so wide of the mark.

These wrecks, upon which the British Government had paid out millions in insurance, were the property of the State, but the chances of raising them were accounted so slight that it was not considered policy to spend further money on them. Well-known salvage concerns, however, had no difficulty in obtaining permission to salve any ship which they had a fancy to raise. They had but to go to the shipping department concerned in order to win a sympathetic hearing. The terms of the contract were on the "no cure, no pay" principle, which meant that any salvage firm with the courage to risk a few thousand pounds in trying to raise a particular wreck was quite at liberty to do so. In return for the concession to work on the wreck, they agreed to give the Government a certain percentage of the value recovered, the percentage being arrived at by mutual agreement. All risk was consequently borne by the salvage concerns, who lost their money in the event of failure and shared their winnings with the Government if they were successful.

The high cost of shipping at that period led to considerable activity on the part of salvage concerns, for if luck happened to be with them there was the prospect of making a fortune out of one operation. But a shipping slump without precedent in all history quickly worked a tremendous revolution. Some new ships halved their value in six months, second-hand ships fell in price from £30 a ton to £7 or less a ton. One great shipping firm had to set aside a fund of half a million in order to write down the value of their new ships directly they were launched, for their new liners were worth more on the stocks than they were in the water. The only way of making their ships pay at all was to decrease their cost, and this could only be done by sacrificing the money saved and placed in reserve. In many cases shipowners paid huge sums to shipbuilders in order to be released from contracts, for they were able to buy new ships at half the price similar ships would cost to build.

This remarkable change was brought about by the great shipbuilding programmes forced on the Allies by the submarine campaign. Not until after the war was the full force of these programmes felt. The new ships coming off the stocks made up the lost tonnage in a few months. The seized German ships helped to increase the slump, and the world found itself richer by 11,000,000 tons of shipping than it had been in 1914. The war had destroyed the markets, the Continental nations had no longer any money with which to buy goods, and the result was the most dramatic change in history. Shipowners who a year previously had been clamouring for ships at any price, were compelled to let 8,000,000 tons of shipping lie idle.

Of course these conditions played havoc with salvage concerns. The fortunes that might have been locked up in war wrecks quietly vanished. It must be borne in mind that enemy torpedoes in the first place had done enormous damage to the sunken ships, and what the torpedoes had left undone the storms of the Armistice years had finished.

The immersion of a ship for a year or two in the sea, with the consequent rust set up in the metal, works sorry havoc, while sand and mud swirling about in the engine-rooms tend not to improve the engines. Every hour that a ship spends on the ocean bed she deteriorates in value. Mud is silting into her, sand and rust are gnawing away at her, the swell is shaking her continuously. The sea soon finds out the weak spots and hammers at them until the whole structure collapses into a fantastic mass. It can be imagined what some of the war wrecks were like after a thousand days of such treatment. They were not worth salving, for no salvage concern would risk thousands of pounds just to recover a little scrap metal. These factors eventually led to a cessation of salvage activity around our shores.

For long after the Admiralty Salvage Section had ceased to operate in home waters, one or two units were working on the Belgian coast, struggling to clear the harbours of Ostend and Zeebrugge from the ships that British sailors had so gallantly sunk in order to prevent the Germans from using them as submarine bases. When the *Vindictive* went down in her allotted place, she covered the British Navy with glory. All the might of Germany, all the skill of which she boasted, failed to move the sunken ship from the spot where the British had placed her. The Germans did their uttermost—for they were anxious to use the harbour—but they were beaten.

The genius of the Admiralty Salvage Section, Commodore Sir Frederick Young, studied the problem. The *Vindictive* was not only full of cement, which had set hard directly the water ran into it, but there were also many mines aboard, and no one knew whether all these mines had gone off or whether some of them were still alive. Added to the problem of the *Vindictive* was the fact that the Germans, in their retreat, had sunk all sorts of craft in the harbour to bottle it up completely, and ensure that the Belgians would never use Ostend again without going to an awful amount of trouble.

For months the divers of the Salvage Section were struggling with the wrecks in Ostend, clearing the channel, blowing tons of cement out of the *Vindictive* in order to lighten her, cutting away hundreds of tons of steel so that there should be so much the less to lift. Mighty steel cables were passed under the *Vindictive* by divers and attached to two lifting craft, one on either side of the ship; two giant pontoons were sunk into place and attached to the hull so that when the time came they could be pumped out and their power used to help lift the stricken ship off the bottom. Some of the compartments in the wreck were made watertight, and after about a year of strenuous toil the task of lifting the structure was undertaken. Pumps were set going, and as the tide rose the shattered British warship came off the bottom and was moved some distance before the falling tide baulked further endeavours. The next day saw the operations carried to a successful conclusion amid scenes of wildest enthusiasm.

The raising of the *Vindictive* signalized the last days of the Naval Salvage Section, but it was by no means the least of the many triumphs that crowned it during the war.

# CHAPTER IX

During the days of the fateful German submarine campaign, the divers of the Admiralty Salvage Section played their part in many a drama, ferreting out clues of vital importance, acting as detectives of the deep. While the *Untersee* boats of the Germans menaced our national existence and ruthlessly committed many crimes against humanity, the deep-sea detectives of the Salvage Section were always on their track, studying their habits, learning their methods, recovering from watery fastnesses those sealed orders which Tirpitz and his staff would have given anything to keep out of the hands of our alert Admiralty.

More than one U-boat, struggling frantically to free herself from the mighty nets in which she had become entangled, found herself caught in a trap from which there was no escaping. The guardians of the nets, going their rounds, marked the agitation of the buoys which told of a giant fish struggling below, and if the prize could not be brought up and captured, a depth charge soon put an end to its struggles.

Sometimes a submarine was found on the bottom without any visible damage to the hull. An accident to her machinery had rendered her helpless. The Germans fought desperately to put things right. As they grappled with the damaged machinery, they saw death coming nearer and nearer. When it was obvious that they could do nothing, that there was no escape for them, many shot themselves to put an end to their sufferings. Entering these steel tombs, the divers of the Admiralty saw ghastly sights—shot Germans lying about all over the place. In some cases it was apparent that the trapped men had been driven mad by their terror and had run amuck and fought each other savagely before they died. They were pitiless to others, but in the end the fear of death had turned their brains and transformed them into madmen.

Of all the submarine crimes which dishonoured the name of Germany, one of the worst was the atrocity of the *Belgian Prince*. It started with the sound of guns and the whine of shells from which it was impossible to flee, and as the wireless mast of the *Belgian Prince* went overboard her captain rang down to the engine-room and the ship heaved to. The U.44 approached warily, waiting to strike again at the least sign of resistance, but seeing that the *Belgian Prince* had frankly surrendered a collapsible boat put out from the submarine, which was now lying idly on the surface, and pulled off to the steamer. Captain and crew of the steamer were ordered to take to their boats and pull to the submarine, and, as they rowed to the U.44 under armed escort, the Germans went down below to open the sea-cocks of the vessel and place bombs to blow the bottom out of her.

Their work completed, the boarding party of Germans rowed back to the U.44. Paul Wagenfuhr, the German captain, ordered the crew of the *Belgian Prince* to line up on the deck of the submarine. They were searched for arms, ordered to take their outer clothes off, their lifebelts were taken from them, and their boats destroyed with axes. Leaving the seamen partially undressed still standing on the deck, the Germans entered the conning tower of their boat and shut it after them.

The crew of the *Belgian Prince* still stood as they were ordered, wondering what was going to happen to them, expecting that now their ship and boats had been destroyed the Germans would take them into the submarine.

Gradually the U.44 began to move on the surface of the sea, and continued to forge ahead for about ten minutes. Then suddenly, without warning, just as darkness descended, the submarine dived, and the forty-three helpless and defenceless men were thrown into the water. For a time the air was rent with their cries as they fought the eternal sea for their lives. Then the darkness blotted out the sights and sounds, and one by one they sank.

It was as deliberate and cold-blooded a murder as was ever committed—the very epitome of that order of the German Naval authorities to "destroy without trace." The destruction of the boats with axes to cut off all means of escape, the deliberate taking away of the lifebelts, the search for weapons, the order to the men to take off their outer clothes, all were thought out, were part of a settled policy on the part of Captain Wagenfuhr, if not on the part of the German Higher Command. All were easy to understand. Even the object of depriving the crew of their clothes, which is obscure to many, becomes plainer upon consideration. Men carry papers and things in their pockets which lead to identification. In taking their clothes from the men, the Germans were also robbing them of their identity, for if any of the poor victims happened to be found clad only in their shirts floating dead in the sea, there was practically nothing to furnish a clue as to who they were, what ship they belonged to, if they belonged to a ship at all.

But the Germans, in their hurried search of the men, overlooked the fact that three of them wore lifebelts concealed beneath their clothing, and these three men, by the aid of their lifebelts, managed to survive until they were picked up. So the world learned of the German crime. But for these three witnesses, nothing would have been known except that the *Belgian Prince* had vanished with every soul aboard.

Throughout August 1, 1917, the naval craft were scouring the neighbourhood for a sign of the U-boat, trying to get on its track. The sea was empty. Casting farther and farther afield, one of our torpedo boats sighted a periscope on the afternoon of the next day nearly a hundred miles from the scene of the outrage. Keen eyes at the other end of the periscope must have detected the torpedo boat almost as soon as the torpedo boat saw the periscope, for our naval gunners had time to get in only a couple of rounds before the periscope disappeared. Racing to the spot, the torpedo boat dropped a depth charge. But she was too late: the enemy was gone.

A torpedo fired at a cattle boat proceeding from Ireland to England furnished the next clue to the enemy submarine. The torpedo missed, and the cattle boat, calling up patrol boats by wireless, managed to escape.

The U-boat hunted warily, for Paul Wagenfuhr had a

definite mission to perform. His task was to lay a minefield in the way of the cattle boats coming out of Waterford harbour in order to interfere with the regular traffic to England. The submarine was equipped with a number of huge mines and special mine-laying apparatus which enabled her to lay these death-dealers while she herself was snugly out of sight beneath the surface. Mostly the mine-laying was done at night, and regularly about once a month a U-boat would scatter her deadly cargo and pen the shipping in harbour until the mines were swept up and a passage cleared.

Hardly a ripple stirred the sea when darkness stole down over Waterford on the evening of August 4. The fisherfolk along the coast, gathering in the village inn, spent an hour or two smoking and chatting over the doings of the day. Some were still standing before the doors of their cottages about midnight when they were startled by the sound of a terrific explosion at sea, a sound that reverberated over the water in the absolute silence of the night. Then, faintly, cries were heard.

The cries sent the fishermen speeding to the quay. In a short time three fishing boats were speeding over the sea, heading in the direction whence the cries came. None knew what lay ahead of them, none troubled even to ask. Death might be lurking for them, but that aspect of the case did not concern them. The sound of the explosion and the cries still rang in their ears, betokening a disaster which sent the fishermen on their swift errand of mercy to succour whomsoever they could find.

Standing alert in the prows of their boats, the fishermen scanned the sea for signs of wreckage. From time to time they called, and listened vainly for an answer. They were about 4 miles from shore when a dark object loomed in the water, a faint cry answered their calls. A minute later a man was dragged over the side of one of the boats.

The stranger was in a bad state. It was obvious he could not long survive. Heading about, the fishermen landed the man as quickly as possible, but stimulants liberally administered had little effect. Just for a time he rallied and managed to gasp out the information that he was a member of the crew of the U.44, and that they were laying mines when a tremendous explosion occurred and shot him up to the surface. His end came suddenly soon afterwards.

The U.44, laying mines in the stilly night to deal death and destruction to others, strayed unwittingly into one of our minefields. One of her mines in floating upwards after its release knocked against one of ours, and the two exploded with such terrible force that the stern of the submarine was practically blown away and the men who manned her were drowned like rats in a trap. Thus Nemesis overtook the Germans.

By Monday, August 6, Commander G. Davis of the Admiralty Salvage Section was recalled from another salvage case with instructions to recover the sunken U-boat. All that night the salvage officer and his men laboured at getting the necessary gear aboard the salvage ship, and at midnight on the Tuesday they reached Waterford.

Early next day minesweepers were at work clearing a passage for the salvage vessel. It was dangerous to move in that area at all, as was manifested during the morning when one of the minesweepers herself struck a mine and foundered. Without waste of time, Commander Davis tackled and raised the minesweeper as a preliminary to the important task of raising the U-boat.

The usual method of finding the wreck by dragging the seabed with grapnels was adopted, and the submarine was located in 90 feet of water, lying right athwart the current which, owing to its strength in this spot, did much to hamper future operations.

The Admiralty was particularly anxious to recover not only the papers of the submarine, but also the submarine itself. Given the German submarine, the British naval experts could go over it at their leisure, see exactly how German design was developing, browse among the latest German improvements and pick to pieces all the most recent German ideas. Not that the British Admiralty lagged behind German design, but it had the good sense not to despise the enemy and to realize it might be possible to learn something even from Germans.

To issue an order for the sunken submarine to be brought into harbour was easy. A few words in code tapped out on the wireless and the thing was done. But the carrying out of the order was beset with difficulties. Commander Davis decided to adopt one of the best known methods of raising the wreck by utilizing the lift of the tide to accomplish his purpose.

One of the outstanding things about salvage experts is their uncanny ability for seizing on any power that happens to be handy and compelling it to serve their own ends. There is unlimited power in the rise and fall of the tides, and the salvage men are clever enough to harness this power to raise wrecks off the seabed. They literally use the sea to rob the sea of its prey, and the ways they follow are more or less those put into practice by Commander Davis, who decided to lift the submarine in a cradle of cables and carry her ashore.

A mighty steel cable was taken from one salvage boat to another, an end was secured on each boat, and the cable was dropped until the loop of it dragged on the bottom. Then this cable was swept under the submarine and hauled along by the salvage boats until they had dragged it into position right under the wreck. Directly it was in place, the two ends were buoyed, and the salvage men began juggling with another cable. One by one the cables were worked into position, and by the ninth day the salvage officer had as many cables as he desired lying snugly under the U-boat from end to end.

The tenth day brought a gale that made further salvage operations impossible. Dirty weather continued for twenty-four days before the gale blew itself out. The salvors, desperately anxious as they were to get on with the job, had perforce to cool their heels ashore while the seas played battledore and shuttlecock with the buoys at the ends of the cables.

On September 10, however, the day dawned fine, and soon after daylight the sweepers were clearing a passage out to the wreck—a task they had to perform every day any work was undertaken. No sooner was the passage swept than the salvors brought to the spot one of those modern lifting vessels which helped to perform many wonderful feats during the war.

In appearance the lifting craft is like a huge, flat barge with a covered deck. Its hull contains a series of great tanks, or watertight compartments, which can quickly be flooded or emptied, just as the salvage expert desires. As the tanks are flooded, so the craft sinks lower and lower in the water, and as they are pumped out so she rises again. When the tanks are full, the lifting craft sits 4½ feet lower in the water, and if she is then attached to a wreck and her tanks be emptied she is capable of lifting a weight of 1200 tons from the seabed.

In raising the U-44 and carrying her to port, Commander Davis, R.N.R., the nearest figure on the lifting vessel, accomplished a brilliant feat. The photograph shows the u-boat just after she was brought to port and also gives an excellent idea of what a lifting vessel looks like.

Say that the difference between low tide and high tide is 16 feet. If the lifting craft be placed in position over a wreck at low tide and pumped out, the cables between the lifting craft and the wreck being made taut, as the tide rises, so the lifting craft swings the wreck off the seabed, and at high tide the wreck lies slung under the lifting craft over 20 feet from the bottom. She can then be towed inshore until she grounds again.

In other words, a vessel floating on the surface is nearest to a submerged wreck at low water. If the tide happen to rise and fall 20 feet, the vessel will be 20 feet nearer the wreck at low tide than at high tide. By filling their lifting craft with water the salvors can bring it another 4½ feet nearer the wreck, and if they then pump out the water tanks they can raise the wreck 24½ feet from the bottom at the top of the tide, provided they have craft capable of lifting a weight as great as that of the wreck. Towing into shallower water follows, as before described.

Hoisting out the deadly cargo of mines from the U-44

Commander Davis placed his lifting vessel in position exactly over the wrecked submarine, and the cables running under the wreck were brought up on each side of the surface craft and securely fastened. The tanks of the lifting craft were blown out with compressed air and, as the tide began to rise, the lifting craft rose with it and dragged the U-boat from her bed 90 feet below the surface. Just before the tide was at the full the salvors began to tow the lifting craft with her burden inshore and succeeded in covering a distance of three-quarters of a mile before the submarine grounded again. Next day, at the top of the tide, the performance was repeated, and the wreck was carried inshore for another three-quarters of a mile. In two days the salvors thus gained a mile and a half, and the wreck now rested on the bottom, about three miles from the beach.

The salvors, making the most of favourable conditions after their enforced idleness, were toiling until far into the night on the wreck. They feared a recurrence of bad weather, and their fears were well-founded. Wednesday brought in its train a strong wind that increased in strength all the morning and made work impossible. By the afternoon it was blowing a gale, and so severe was the storm that one of the salvage lighters was unable to withstand its fury. She started to founder, and it was only with the utmost difficulty and in the face of tremendous risk that one of the salvage men managed to get aboard and bring her safely to harbour.

The calm courage and confidence of the salvors were things to marvel at. They knew beyond doubt that live mines were aboard, and that these mines were liable to go off at the slightest jar and blow them all to pieces, yet they went about their jobs for hour after hour, day after day, as though such things as mines did not exist. Time after time the sea bumped the submarine against the bottom and, every time it happened, death in its most horrible form hovered near them. Once the submarine dropped sheer from the cables, and no one knows even now why they were not all wiped off the face of the sea. There was just one tense moment, then, as nothing happened and their luck held good, they started to get the submarine back into the slings again.

Another lifting craft was brought on the scene and, picking up the wreck again, the salvors went ahead with the work tide by tide. In their passage shorewards they performed the extraordinary feat of carrying the wreck over a bar of sand that rose steeply for 14 feet—an operation requiring the greatest skill and delicacy in adjusting the lifting cables. The nose of the submarine had to be lifted inch by inch until it attained an angle that enabled it to rise up the slope without digging its bow into the sand. Had the nose of the craft been lifted too high, she might easily have slipped backward out of the cables supporting her, and such a slip might not have ended so happily as the previous one. However, Commander Davis succeeded in negotiating this supreme difficulty surely and safely, and his brilliant work was later rewarded with the Distinguished Service Cross.

In the end, after making twenty-one lifts in twenty days, the salvors beached the infamous U.44. She proved a golden haul, for the mass of confidential information recovered from her turned out to be of the utmost importance. She had on board nine mines, which were cautiously taken out by Commander Davis and rendered innocuous, besides several torpedoes and a big collection of shells.

Followed the grim and ghastly task of disinterring the

dead. On September 26, twenty-one bodies were removed under the direction of a surgeon and carefully searched. One by one the dead Germans were sewn in canvas and weighted with firebars.

That evening the salvage ship, fitted for the occasion with special platforms on which the bodies were placed, steamed out to sea. At midnight she stopped. The salvage men with bared heads stood solemnly by while the chaplain read the burial service in grave, sonorous tones. Then, very reverently, the dead were committed to the deep and the cleansing sea closed over them.

# CHAPTER X

Although we live in an enlightened age, superstition is still rife, and not many people would care to dive for the first time in a submarine bearing the unlucky number 13. Yet in spite of the fact that sailors are generally credited with being more superstitious than most people, no thought of danger crossed the minds of the seventy-three men who during the war stepped aboard the British submarine K.13 in order to carry out her trials. She was a wonderful craft, 334 feet long, just under 27 feet wide amidships, and as she lay at her moorings she displaced 1880 tons.

Like her sister ships of the same type, she was one of the fastest submarines afloat, capable on the surface of overtaking most battleships in order to send them to their doom, able to take her place with the Grand Fleet and steam along with them at top speed without being left behind. This wonderful speed was attained by fitting her with steam turbines in addition to the usual oil engines and electric motors. Her stumpy funnels folded down when she was diving, and the introduction of steam made it essential to fit fairly big ventilators. In order to dive she could take into her ballast tanks 800 tons of water in four minutes, but with a big submarine over 100 yards long, all divided into many compartments, diving was a delicate operation that depended for its safety upon all the men carrying out their duties instantly. It was necessary that the crew should be quite conversant with their craft and that there should be perfect team work. But an absolutely new craft is bound to present some strange features to her first crew. In this case she was a new development in submarine practice, and it was probably the fact that the K.13 was unfamiliar that brought about the ensuing disaster.

Built on the Clyde, she was taken along to the Gareloch to be put through her paces. The Gareloch was quiet, away from spying eyes, free of the attentions of the unwelcome enemy submarine, and here the K.13 carried out her surface trials satisfactorily. The conning tower was closed, the funnels were dropped back flush with the deck, and orders were given to trim the boat for diving. The watertight doors were shut and the sea began to flow into the tanks. Then, as the craft submerged, came disaster. A mighty rush of water swept into the after part of the ship, drowning instantly the thirty-one men on duty there, and carrying the K.13 stern downwards to the bottom. It was afterwards discovered that in diving some of the ventilating scuttles had been left open and these had flooded the stern of the ship. It was a tragic oversight that in a moment swept thirty-one men into eternity.

In the forward part of the K.13 forty-two men were imprisoned, held fast on the seabed by the weight of water in the ship. There was no trace of panic. Nobody turned a hair. As quietly as though they still floated serenely on the surface, they stood by and carried out their commander's orders.

For hours they strove to get the ship to move, to lighten the tanks sufficiently to bring her to the surface again. The ship remained fast. No trace of movement was to be detected. The watertight bulkhead across the centre of the vessel held death at bay for the moment, but no one knew how long it could withstand the terrific pressure. At the other side of the bulkhead lay their dead companions, and the hungry sea was waiting to engulf the living. Death threatened them from all quarters, death from drowning, death from asphyxiation owing to the exhaustion of their air supply, death from starvation even if the air held out. Hour by hour death came nearer. They realized it only too well, but still they remained cheerful.

When it was seen that all their efforts were useless, Commander Godfrey Herbert, D.S.O., who was in command, and Commander F. H. M. Goodhart, D.S.O., who was aboard to watch the behaviour of the vessel before taking over the command of K.14, conferred and agreed to try to get to the surface, 90 feet above their heads, in order to obtain help. They knew perfectly well that they were probably going to their deaths, that the odds were so tremendously against them that they were not worth considering. They did not think of themselves; they thought only of the forty men caught in that death-trap.

The one way of getting to the surface was through the conning tower. But the terrific weight of the water above closed the lid so tightly that the strongest giant in the world could never lift it. To raise it were beyond the strength of mere human beings. The only way of accomplishing the feat was to let into the conning tower compressed air until the pressure of the air equalled the pressure of the sea, and as the air burst a way upwards the gallant officers hoped to be carried with it to the surface.

Quietly they entered the conning tower, and partially flooded it. The compressed air was turned on. Minute by minute the pressure increased, minute by minute the officers waited, wondering if death or life was to be theirs, whether their attempt was to succeed or fail.

So great grew the pressure that the air could no longer be kept within bounds. With incredible strength it burst upwards and Commander Goodhart was dashed violently against the steel sides of the conning tower and killed instantly.

By the greatest good fortune Commander Herbert missed the full force of that deadly upthrust of air. Still he, too, was hurled upwards and, as the water rushed in and the air gushed out, was carried clean through the conning tower to the surface.

Already the disappearance of K.13 was arousing anxiety up above, and a salvage craft had been called to the spot. A couple of men in a boat, noticing the figure of Commander Herbert as he came up in the Gareloch, pulled quickly towards him and dragged him over the side. He was almost dead with exhaustion, and the wonder is that he ever survived that terrible ordeal.

As soon as he was sufficiently recovered, he gave an account of what had happened and told how the men were trapped in the submarine. The urgency of the case was obvious. It needed no stressing.

Then began one of the most thrilling salvage fights in the history of the human race. It was a fight, not for treasure, but for human life. It was a race against time, a long tussle with death.

Divers dropped down the shot-ropes to the bed of the Gareloch and began to search for the sunken submarine. The light was none too good, owing to the water being fogged with mud, but they were searching only a short time when the dark hull of the submarine loomed in front of them. They hurried up to it. One drew an axe from his belt, hammered hard at the side.

Answering knocks came from within, and those waiting anxiously on the surface heaved a sigh of relief as the divers telephoned up:

"We've found her. They're still alive!"

Surveying the wreck, the divers discovered that the bow of the submarine was about 20 feet higher than the stern, which was already covered by a dozen feet of mud. Wading in slime sometimes up to the armpits, the divers worked their way round her, then quickly sped to the surface and reported her position.

At once the experts summed up the situation. The K.13 with her stern full of water, covered up aft by a dozen feet of mud, was too heavy to raise bodily. She was well over 3000 tons, and up to that time nothing like this weight had ever been lifted from the seabed. The only thing to be done, the sole hope of saving the imprisoned men, was to strive to lift the nose of the craft to the surface while leaving the stern resting on the bottom. Nothing else was possible.

"The first thing to do is to get through supplies of food and air to them," the salvage officer remarked.

The divers slid down to the bottom and, disregarding all thought of their own safety, laboured hard and long to connect up with the entombed men. They must have broken the endurance record of the world, for one worked for over twelve hours continuously on the seabed without taking food, without resting. Time was too precious for them to waste a second. They realized the risk, but they accepted it as gladly as Commander Goodhart ran the risk which led to his death. They worked until they were ill and dizzy, floundering in the mud, wrestling with giant steel cables.

Forty men were depending on them for their lives. The thought nerved the divers to prodigious things. It was essential to communicate with the imprisoned men, to let them know that everything possible was being done for them, to strive to sustain their spirits. Commander Kay of the Salvage Section found the way. Sending down a submarine flash lamp, he instructed the divers to rig it up in front of the periscope. By peering into this instrument the prisoners were thus able to read the messages that were flashed to them in Morse Code, and were made to understand that they were not entirely cut off from the world after all. With many a struggle, the divers managed to open a valve in the hull and to attach a pipe through which food such as Bovril, bottles of hot soup and chocolate, as well as life-giving air, were passed from the surface. All this entailed long hours of endeavour.

The coolness of the men in the submarine was almost unbelievable.

"Send us down a pack of cards to while away the time!" one shouted up the pipe.

The cards were procured and sent down, and these British seamen played cards while Death peeped over their shoulders.

Up to then the men had been carefully conserving their supplies of compressed air, not knowing how long they would need them to keep alive. Now that air was being pumped from the surface, they were able to use what was left of their own supplies to blow all the oil out of the forward tanks. This lightened their craft considerably.

After a terrific struggle, the divers managed to fix mighty steel cables under the nose of the submarine. Salvage craft and lifting vessels strained away. For a time they made no impression. Then slowly the grip of the mud began to relax and the bow of the submarine, lightened by the blowing out of the oil tanks, began to rise nearer and nearer the surface until, about midnight, it broke clear into view.

It was a weird sight. Great arc lamps lit the scene, and under their glare the salvage men attacked the steel hull of the K.13 with oxy-acetylene blow-pipes. Every one was desperately anxious, afraid that the submarine might slip. Under the intense heat of the blow-pipes, the steel grew soft and melted. Gradually, laboriously, the salvors burned their way through the stout outer plates.

History reveals no more thrilling rescue than that of the survivors of the K.13 after she had been at the bottom for two and a half days. This rare photograph shows the bow of the K.13 after it had been hauled to the surface to enable the men to be cut out.

They now made an onslaught on the inner hull, directing the flame on the steel shell. The metal glowed and flowed. A rush of air leaped upwards from the interior of the vessel and blew out the roaring flame of the blow-pipe.

"Get us some matches!" the divers called to those above.

Under their very noses a hand from inside the ship suddenly slid through the hole in the metal, the fingers holding up a box of matches.

"Here you are," said a cheery voice, and the divers knew that all was well.

Another period of strenuous endeavour and the hole in the metal was big enough for a man to squeeze through. Then, as the forty prisoners were helped and carried to freedom, the cheers of the salvage men echoed to the shore.

Never will men be nearer death than those saved from the K.13. For fifty-seven hours they were imprisoned in the sunken submarine at the bottom of the sea, for two and a half days they lived with death at their elbows, not knowing when the end would come. Their ordeal has never been equalled, and their rescue is one of the most thrilling deeds in the annals of sea salvage.

Barely were they rescued when a storm arose. The cables holding up the K.13 snapped asunder, and the submarine plunged again to the bottom. The men had been cut out not a moment too soon.

In due course followed the salvage of the unlucky K.13. It was effected solely by the use of compressed air, which was pumped down one pipe into a compartment until it had driven all the water away through another pipe to the surface. In this way she was pumped out compartment by compartment, but even when all the water was expelled she still stuck in the mud. For two or three days the salvors strove to drag her from the clinging mud, but not until she was freed of the overlying silt by sand-pumps did she bob to the surface just like a cork. Proving little the worse for her adventure, she was put into commission again under another number, so the unlucky K.13 vanished for ever from the British Naval Lists.

# CHAPTER XI

Quite as thrilling as the experience of the men who went down in the K.13 was the adventure which befell the crew of an American submarine, the S.5, and it is doubtful if any popular novelist, with all his imagination and powers of invention, ever thought out a more remarkable situation than that in which these American sailors found themselves.

The American submarine concerned had been travelling on the surface, when the commander gave the order to prepare to dive. Down she went, and for a time glided unseen in the depths. Then her commander got ready to bring her up once more.

Of a sudden something went wrong. The air failed to blow out the forward tanks. The men felt the floor slip away under their feet as they rose. They were thrown on their backs, on their faces, rolling sideways in all directions. There was no shock, not the slightest jar. The submarine just swung like a pendulum, and when the officers and men managed to disentangle themselves from the various positions into which they had been thrown, they found the bulkheads had changed places with the floor of their craft.

The submarine was actually hanging perpendicularly, bow downward, with just the end of the stern showing above the surface. It was a terrible plight to be in, and every man aboard recognized at once that he was face to face with death. Their only hope was that a vessel would sight them and manage to rescue them before their air gave out, yet there was so little of the stern peeping above the surface of the sea that the odds against it being noticed were tremendous.

Most submarines nowadays are equipped with a portable telephone which can be floated to the surface, where it is supported by a buoy. This telephone was designed for just such an emergency, and the commander quickly uncoiled the cable and sent the telephone floating upward.

Followed a most nerve-racking experience. For hour after hour they swung about under the sea, rocking this way and that, spinning sometimes like a top, ringing on the telephone at regular intervals, and waiting tensely for the sound of a voice to tell them that they were found. All day they waited without any reply. Air was being used up every minute, and death by suffocation was not pleasant to think upon. Even worse was the thought that at any moment the submarine might cease to swing, and would plunge to the bottom like a stone, fracture her plates and wipe them all out in a few seconds.

Twenty-four hours passed. All through the darkness of night until dawn those insistent signals went up to the telephone and a sailor waited tensely for an answering voice. None came.

Another day of suspense began. The men were like prisoners in a condemned cell, not knowing whether they were going to their doom or whether a reprieve was coming. All the time they were striving to find out what was wrong, struggling to right their craft again. The task was beyond them. Their efforts were of no avail. Still they rocked and swung like a pendulum in the broad Atlantic. It was a nightmare situation. For men to remain so strong and yet so helpless was maddening. So the dreadful hours crept by.

An American transport, the *General Goethals*, was steaming down to Panama when one of the men aboard thought he heard the sound of a telephone bell.

"What's that?" he said.

His companion looked at him, "What?"

"Sounded like a telephone," said the first man.

His shipmate was about to retort when he, too, heard the sound of the bell.

"There it is again," said the first man.

"Sure!" answered the companion.

Other men came crowding up.

"What's wrong?" they inquired.

"Didn't you hear it?" asked the first man.

"What?"

"The telephone!"

At that moment the sound came to them again. They

looked at each other. Some wondered if they were bewitched. They were far out on the open sea, and it seemed impossible that a telephone bell could be ringing there.

More and more men crowded round, and more and more heard the bell. There was no mistaking it. It was certainly a telephone bell. So plain was it, so insistent, that at last the captain signalled down to stop the engines.

Half a dozen seamen took their places in a boat. Outwards it swung from the side of the ship and a moment later sat with a splash in the sea. Rowing in the direction of the mysterious sound, the sailors at last sighted the buoy with the telephone attached. The stern of the craft was barely visible.

Imagine the transports of those unfortunates when voices hailed them cheerfully from above! They had been swinging about in their awful predicament for thirty-five hours when the telephone was picked up, and air was running so short that they had only enough to last them for an hour or two longer.

Instantly the men below made clear their peril. The troopship flashed out her wireless call for help.

Not a ship within radius heard the call.

Then cropped up another of those strange tricks of Fate. An American schoolboy, named Moore, keen on wireless long before the wireless boom set in, was experimenting with his home-made set when he picked up the call. Proudly he sent out this message of life and death on his own transmitter. The nearest naval depot picked it up and destroyers with special plant aboard were hurried at full speed to the rescue.

Meanwhile the captain of the transport had managed with the greatest difficulty to get strong hawsers round the submarine, lashing them tightly to his transport in order to keep the stern of the submarine above water. Then his engineers after a deal of labour cut a small hole in the steel skin and began to pump fresh air in to the prisoners.

This was the situation when the destroyers appeared on the scene. Immediately they fixed more hawsers round the submarine to prevent her from slipping to the bottom, and with the special appliances at their command they managed to cut through the rivets and force out one of the plates of the up-ended craft.

One by one the twenty-seven men and their commanding officer scrambled through to the open air again, after being imprisoned for forty hours in that crazy submarine swinging about under the sea. Thus a telephone ringing in the open sea, where no telephone could possibly be expected, and a boy playing with his wireless set were instrumental in saving the lives of an entire crew after a most terrible experience.

Not so fortunate were the crew of a British submarine which, like the K.13, met with a mishap that sent her plunging to the bottom. All were killed except one man, who with his own lips afterwards related how he had battled with death and won his way back to life after one of the most amazing adventures that have ever befallen man.

He happened to be in the engine-room when he perceived the water pouring in through the conning tower in one mighty cascade. In a flash he realized that the boat was doomed. Rushing along the engine-room he shouted at the top of his voice to warn his comrades in the other parts of the ship. The sea swept into the engine-room after him. In a moment the floor was flooded.

Fast as he moved, the water was faster. Before he could get out, he heard the sinister sound of the engine-room door slamming. He turned and thrust his shoulder against it. It would not budge. He was trapped in the engine-room of a sunken submarine! The rush of water had closed the bulkhead door, and the space beyond was completely flooded, making it impossible for the imprisoned man to move the door. Even if he had succeeded in opening the door, it would have been merely a matter of seconds before the hungry sea drowned him.

He stood to compose his thoughts, to make up his mind what to do. More than once he had imagined himself trapped in just such a manner, and he was well aware that if he could succeed in equalizing the pressure of the air inside with the water outside he might get out of the submarine and escape.

But to work things out in theory is much easier than to carry them out in practice, especially if your life depends on your doing everything exactly as it should be done, when the least little slip means death.

The man reached out his hand to grasp a metal lever. His

fingers closed on it. He recoiled from a severe electric shock. He touched something else, and again felt the jolt of electricity. His knee knocked against one of the engines and he felt a big shock in his leg. Very gingerly he put his finger on another metal object, and once more experienced the sensation of electricity. Everything around him was charged with electricity, and it was some time before he realized that the flooding of the engine-room had short-circuited the electric current.

Now another factor crept in to make the situation still more desperate. The sea water, flooding the electric batteries, began to set free chlorine gas. The smell of it grew stronger, made him gasp. So to the risks of drowning and suffocating was added the danger of gas poisoning.

In like circumstances few men could have kept their nerve. Most men would have abandoned themselves to their fate, would have given up all hope in the face of so many perils. But not this British sailor. With all his strength he began to fight to get out of the submarine, to put his theories into practice in order to save his life. He must have possessed tremendous will power, wonderful courage and determination.

He tried the torpedo hatch, to make quite sure that the pressure above was such that he could not shift it. He might have been pushing against Mount Everest itself. Wasting no time, he set the bolt of the hatch so that the merest touch would release it, then he opened a valve to let in more water. As the water flooded the compartment, the air in it was compressed more and more. Higher and higher crept the water, greater and greater became the pressure of the air until he felt he could stand it no longer. He slipped the bolt of the hatch, and as he felt it give to the pressure he slipped a hand on the outside. A gust of air swept out, held up the cover momentarily, then the great metal lid slammed down again, crushing all the fingers of the brave man's hand.

Maimed though he was, his courage remained unshaken. Giving up his idea of escaping by raising the air pressure, he determined on the most desperate expedient of all. He made up his mind to flood the compartment completely, when the pressure of the water inside and outside would be equal, and he could open the hatch—if he were not drowned in the attempt.

Opening more valves, he scrambled on top of the engines and watched the water pouring in. It rose to the hatch coamings, till only his face was above the surface. Then with a quick heave of his shoulder he pressed against the hatch. The imprisoned air burst out and the water rushed in, sweeping over his face and head. Holding his breath, he thrust again at the hatch, which luckily passed the vertical and fell backwards with a clang. Then he struck out desperately towards the surface.

A destroyer steaming along saw a tiny patch of white in the water. It was the face of the hero of the submarine. He was to all intents lifeless, practically dead. Wasting not a moment, they forced the water from him and after a hard struggle succeeded in bringing back to life one of the bravest men who ever breathed.

Not without its amusing side was the adventure which befell three unhappy men on an American naval submarine. She was engaged in making a series of cinematograph pictures, and orders were given to prepare for a very rapid dive, known as a crash dive.

Two cinema men were still standing on the deck with their cameras, and the commander was in the top half of the conning tower, which was, of course, open. To their consternation the boat began to submerge. Realizing that there had been some misunderstanding, and thinking only of saving his ship and crew from a terrible disaster, the commander, who had no time to enter the ship, shouted to the men to close the hatch under his feet.

It was slammed not a moment too soon, and the commander inside the conning tower was carried beneath the surface. His first thought was to escape. He scrambled upwards towards the opening. Something stopped him, held him fast, kept him a prisoner.

What had happened was that a projection in the conning tower had caught in his open pocket and was holding him down.

Struggling desperately, and swallowing a deal of water, he managed to tear himself free and kick up to the top.

Gulping in the fresh air, he looked around him. One cinema man was swimming strongly some little distance away. Of the other, there was no trace.

Just as the commander was beginning to give the other man up for lost, the submarine herself reappeared. The commander gazed at her in astonishment, hardly believing his own eyes. With her came the half-drowned cinema man, his arms thrown round his camera and the wireless mast, and clinging to them like grim death.

"What the dickens did you go down with her for?" asked the amazed officer, when he was taken aboard.

"I couldn't swim a stroke, so I thought it safer to stick to the ship," explained the camera man naively.

Luckily for him the crew instantly saw that something was wrong and brought the boat up at once.

So recently as the last days of October, 1923, two American seamen, Henry Breault and Lawrence Brown, were immured for thirty hours in a submarine at the bottom of a bay near the Panama Canal. Breault most heroically dashed into the ship as she was sinking to see if he could assist anybody who happened to be within. He found Brown asleep in the torpedo-room, and they just succeeded in closing the door when the O.5 went down in 40 feet of water.

There was not a morsel of food aboard, not a drop of drinking water. First the lights failed, then the batteries exploded and caused a fire which blazed furiously for some time.

Meantime, a third man, Charles Butler, caught in the engine-room, took refuge in an air pocket, stripped off his clothes and made for the hatchway. Emulating the plucky fellow who escaped from the British submarine, he thrust open the hatch. So enormous was the pressure that he was blown right out of the water, breaking the surface like a leaping salmon. He was soon picked up, after being at the bottom for eight minutes.

In three hours the other two prisoners heard the knocks of a diver and knew that attempts were being made to rescue them. Nine hours later they felt the submarine begin to move upward. For a little time she continued to rise, then their hopes were dashed by a sharp snapping sound and they felt their craft fall with a bump to the bottom again.

The ticking of the clock for hour after hour, the dreadful dragging of the hands round the face of it nearly drove them distracted. They could not bear to watch it longer. There they sat, wondering, hoping.

Another sixteen hours passed before they felt the submarine again begin to rise, moving so slowly that both men were consumed with anxiety. The maddening clock ticked on as the craft was wound up. Water splashed on the deck, the pent-up air gushed out, footsteps sounded and they knew deliverance was at hand. Breault pushed open the hatch and both men stood blinking blindly in the dazzling sunshine.

Their heads reeled. So sick and ill were they owing to the sudden change of pressure that grave danger was only averted by quickly placing them under the same pressure in another submarine, and then slowly reducing the pressure in accordance with the recognized diving practice. Thus they came unscathed through their dreadful trial.

The K.5 during battle practice with the British Fleet in 1921 sank in such deep water that no attempt was made to recover her. But the American naval experts, when a similar disaster overtook the submarine F.4 at Honolulu in March, 1915, were so anxious to find out what had happened that they determined to do their utmost to retrieve the sunken craft.

Going out for a practice spin, the F.4 quietly submerged and was never seen again. Boats were soon in search of her, and the result of dragging operations led to her discovery on the bottom outside Honolulu harbour in just over 50 fathoms, or 304 feet, of water.

Unhesitatingly the greatest salvage experts in the world would have pronounced her lost beyond recovery. She was 100 feet deeper than the British record dive of 210 feet, a depth which no other divers in the world had ever reached, and she was far deeper than any craft hitherto lifted from the seabed.

The experts of the American Navy, aware of these and other facts, knew that they desired to achieve the impossible, but instead of admitting that it could not be done they straightway set about doing it. A big rise and fall in the tide would have been of tremendous assistance to them, but at Honolulu the tide rises and falls only 18 inches. It was of no help to them at all. So they made their plans to haul her up bodily by winches and tow her into shallower water until she grounded; while for the last stage of the journey into the harbour they placed their faith in six pontoons, each sheathed in a jacket of timber 4 inches thick to prevent the cables from cutting it. This stout timber casing successfully protected the pontoons from all damage when they were brought into play. Nor was it unnecessary, for, incredible as it may seen, the chafing of the submarine during a sudden gale quickly wore through the mighty steel cables as she rubbed them against the bottom.

It was in connection with the cables that the greatest diving feat in all history was accomplished. The cables were swept underneath the submarine by surface craft in the usual way. But the salvors could not be sure that the cables were exactly where they ought to be. With cables too near the bow and the stern, the submarine would just fold up as she was lifted and break her back, the two halves, falling apart, probably defying recovery. Even if they could be raised, the damage would be so great that all traces of the original accident would be destroyed and the experts could never learn why the submarine had foundered.

The one way of finding out whether the cables were properly in place was to send down divers to see. A diver in Lake Huron in the 'nineties, trying to recover sunken treasure, was crushed to pulp at a depth of 198 feet; even a diving bell, operating later on the same wreck, was unable to withstand the pressure, consequently it seemed like sentencing a man to death to order him to dive to a depth of 304 feet. However, the cleverest diving expert in the American Navy pondered over the matter and, in the light of recent experiments, considered it could be done provided all the rules were most rigidly observed. The finest divers in the American Navy, men who had been specially trained, were thereupon sent to Honolulu to carry out this gigantic task.

The leading diver struggled into his suit. For aught he knew, he would never come up alive; the enormous pressure of the water might squeeze his unprotected legs and body and arms until it had squeezed all the blood in his body through his eyes and ears and nose and mouth. He knew that the metal helmet protected his head from the sea pressure, which was the reason why the nip of the sea drives all the blood in the body up to the head. But he smiled cheerfully as his helmet was screwed into place.

A few moments later he was sliding down the shot-rope. Down and down he went, the sea pressing heavier and heavier on his body. Up on the surface the air pumps heaved quickly to pass down to him the air that would prevent him from being squeezed to death.

Reaching the wreck at last, he found the pressure so enormous that it was almost impossible for him to lift his hand in the water. To move at all was really like pushing his way through some solid substance. Nevertheless, he managed to survey the wreck and was slowly drawn up again to safety, after spending ten minutes at the bottom.

Several times he and his fellow divers penetrated to these startling depths to see that adjustments were properly made. Then, just when everything seemed all right, the sense of impending tragedy gripped the watchers on the surface. They had drawn up one gallant diver to 200 feet, when he found that his lines were entangled and that he was stuck fast. It was a fearful situation. For a diver to be caught at this great depth is almost certain death.

Relays of divers were sent down to his aid, and for two hours they struggled and fought to release their comrade who was dangling there at death's door 200 feet below the surface of the sea. In the end they disentangled him, and he was drawn up in a most critical state. Double pneumonia struck him down, and for months his life was despaired of. Eventually a fine constitution and tireless nursing enabled him to pull round and regain his lost health. But it was a desperately close shave. That any man could reach this depth and still live is little short of a miracle.

Eventually the ill-fated F.4 was towed into harbour. In raising her according to plan, the American Navy broke three records. By attaining the incredible depth of 304 feet, the American divers wrested the diving record from the British Navy; that unfortunate diver who was forced to remain at 200 feet for two hours, without fatal results or permanent injury, created another record; and their third record was achieved by lifting the submarine from the greatest depth at which any wreck has ever been raised. It is impossible to praise the divers and salvage officers too highly for these magnificent feats.

If the American Navy has robbed the British Navy of the diving record, the British Salvage Section still has a few more records left. For instance, when a German submarine was put down in 190 feet of water off our rocky northern coast, the British Admiralty calmly ordered the Salvage Section to bring the submarine to port.

In the face of a definite order of this sort, there was nothing to be said. The Director of Salvage hastened to the spot, and sent divers down to survey the wreck and if possible recover the papers. They found an arm protruding from the partly-closed conning tower, the fingers, stiffened by death, clutching as in a vice some of the secret orders which the commander was endeavouring to cast away when he saw that capture or destruction was inevitable. Before he could rid himself of the papers, the submarine plunged to her doom and the cover of the conning tower slammed down on his arm.

With an effort, the divers unlocked those clammy fingers and took the papers. Then they managed to raise the lid of the conning tower and enter the ship, although it was practically at the limit of the depth at which divers can possibly work. Their submarine lamps lit the gloom of the interior, and a search brought to light the log and other papers, which were sent post haste to the Admiralty.

The order to take the wreck to port was much more difficult to obey. She was down on such a rocky coast in such a position that lifting her in the ordinary way was quite out of the question. Commodore Young thereupon decided to do what had never been done with a craft of this size since the world began, that is, raise her from the depths by sheer mechanical power. The cables were swept underneath, and divers saw that they were properly in place. Then the powerful machinery installed in the salvage ships began to work, and slowly but surely the great steel cables, thicker than a man's wrist, were wound up until the U-boat was within a few feet of the surface. It was an extraordinary feat to lift this wrecked submarine, weighing nearly 1000 tons—practically four times the weight of the American F.4—from a depth of 190 feet by the sheer power of machinery.

The salvors crowned this remarkable effort by carrying the submarine in her cradle of slings nearly 40 miles round the coast, which was another record the British Salvage Section made that month. Just as they got her to the mouth of the harbour, she slipped from the slings and went to the bottom again. Picking her up once more, the salvage men towed her into dock so that the submarine experts could dissect her.

Another astonishing feat performed by British salvage men was the raising of a collier that sank right in the fairway at Rosyth. The danger of other ships striking her and piling up was so great that her removal became imperative. To pick her up in the approved style by sweeping cables under her and using lifting craft to swing her clear of the bottom was the obvious way of clearing the channel. But she was a dead weight of 3000 tons, or about 1000 tons heavier than the heaviest wreck raised by such methods.

If her cargo had been bales of cotton or something easy to handle, divers would have gone down and removed part of her burden in order to lighten her. But coal is about the worst thing in the world to deal with under water. Consequently the salvors tackled the job with a brace of lifting craft, which enabled them to master 2400 tons, and a couple of mighty pontoons, which provided the power to lift the remainder. Everything was fixed, and as the tide rose the salvors managed to drag the wreck out of the way of other ships, and eventually, after a terrific fight lasting a considerable time, succeeded in beaching her.

Commodore Sir Frederick Young also mastered a weight of about 3000 tons in lifting Captain Fryatt's ship, the *Brussels*, at Ostend, and these two feats performed by British salvage experts constitute a world's record for the greatest deadweight ever raised in recent times from the bottom of the sea.

# CHAPTER XII

The resources of the salvage experts in fighting for the life of a ship are amazing. They will cheerfully run the gravest risks, do the most extraordinary things to get her into port. But that they, whose avowed aim in life is to save ships, should deliberately sink them, savours of something akin to madness. Yet occasions arise when prompt decisions have to be made, when the salvage officer is literally between the devil and the deep sea. An outbreak of fire aboard a ship places him in this quandary. Damage to a ship by water can be remedied, but fire, once it gets a hold, consumes ship and cargo. Of two evils, the salvage man chooses the lesser, and if there is no other way of combating the fire he will calmly sink the ship as a preliminary to saving her.

Giant oil tanker which blazed for days, blotting out the heavens with dense clouds of smoke. The salvage men were eventually compelled to sink her to put out the fire.

More than once during the war British salvage officers had hot times with burning ships, and one of their most thrilling adventures sprang from a collision between two oil tankers called the *War Knight* and the *O. B. Jennings*. A big convoy of ships was proceeding along the English Channel in the early hours of March 24, 1918. It was pitch dark, and the ships with their attendant destroyers were steaming at full speed without lights in order to dodge the attentions of German submarines. Too late the officers on the *War Knight* saw a dark shape appear immediately in their course. A moment afterwards came a terrific impact. The bow of the *War Knight* cut into the side of the *O. B. Jennings*, bursting one of the mighty tanks full of naphtha. It flashed into one gigantic flame which instantly blotted out most of the crew of the *War Knight*, and in a minute or two a Niagara of naphtha from the fractured tank was setting the whole sea ablaze. The one or two men still alive on the flaming *War Knight* frantically hurled themselves overboard, to meet a terrible end in the fiery sea. It was an awful sight.

The fire leaped to the skies, while the men of the *O. B. Jennings*, in that moment's respite before the blazing naphtha floated round to the other side of their ship, rushed to their boats and got away. But Captain Nordstrom and his officers stuck to their ship, though she was belching flames and every moment her other tanks threatened to explode and blow her sky high. Then a British destroyer speeded into the full glare of the light, and one by one the little band of heroes jumped to safety. The captain, leaping last, slipped between the two vessels to what seemed certain death, and for a space it seemed that he, too, was to lose his life, but the prompt measures of the British sailors eventually led to his rescue.

By now the two ships were blazing like funeral pyres in a sea of flames. Great billows of smoke rolled from the stricken tankers in the dawn, blotting out the heavens, looking almost solid enough to stand on. With incredible pluck a naval officer, watching his opportunity, plunged into the inferno aboard the *War Knight* and made fast a mighty steel towing hawser. Jumping back to his ship, he took in tow the flaming tanker which had now drifted right into one of our minefields. It was a gallant piece of work. British mines were all around him, waiting to blow him to pieces, but regardless of danger he kept his course. Once a big explosion shook the stricken vessel as she struck a mine. Luckily, the ship towing her escaped, and the salvage officer, seeing at last that it was not possible to prevent the tanker from burning out, decided to sink her by gunfire on a sandy bottom where there was at least the prospect of salving her later on. Never again, however, did the *War Knight* sail the seas. She proved a total loss.

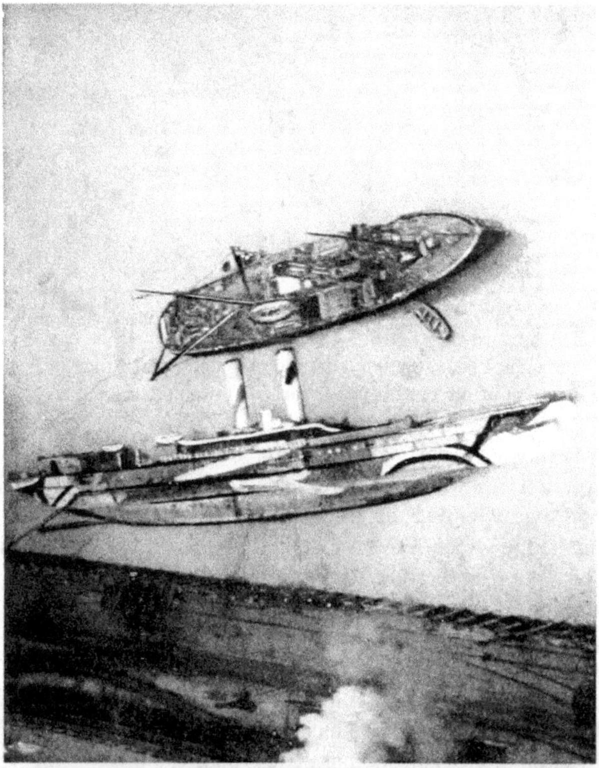

A striking photograph, taken from the air, of the camouflaged troopship onward lying on her side by Folkestone quay after she had been scuttled to put out a fire. The salvage ship is anchored just off the ends of her funnels, while the railway lines on the quay are seen in the foreground, the upright piles of the quay itself having the appearance of the sleepers of a railway track.

The *O. B. Jennings* was also taken in tow and brought to

Sandown Bay in safety. Day after day the fire continued to rage in her, vast clouds of smoke continued to foul the heavens. Nothing could quench the flames, and at the end of ten days the Admiralty salvage officer gave instructions for a torpedo boat to shell the tanker until she sank.

The onward with her funnels cut off and deck houses removed. note one of her propellers just showing above the water and also the lifting craft between her and the salvage steamer.

It was a desperate remedy, but it proved a brilliant solution of the puzzling problem. As she went down, the sea just overwhelmed the fire and allowed the salvage men to tackle the wreck. Divers tapped the undamaged tanks of the ship, pumps were connected up and 8000 tons of oil taken from the sunken vessel. Then the places where the shells had pierced the hull were repaired and the *O. B. Jennings* was pumped out and floated into dock.

A patch was put on her wound, and she set out for the United States; but, as ill-luck would have it, she was caught by another German submarine less than 100 miles from New York and sent to the bottom for good, so all the efforts of the British salvage men were wasted in the end. That collision cost Great Britain just £1,000,000.

Another outstanding case where the ship was deliberately scuttled in order to put out a fire was that of the troopship *Onward*, which carried many thousands of troops to France. She was lying about midnight at the quay at

Folkestone when flames suddenly burst from her, owing, it is thought, to a thermit bomb secreted by a spy. She blazed up furiously, threatening destruction to the whole quay and endangering our communications with France. The destruction of the quay at that time would have been a disaster compared with which the loss of the steamer was as nothing, so quickly the decision was made to sink the *Onward* by opening her sea-cocks. This was done, and the fire went out in a venomous hiss as the sea swept in.

Unluckily, in sinking, the ship turned over on her side, and before she could be raised she had to be set upright. As she lay, she was preventing a much-wanted berth of the quay from being used, so the Salvage Section was given a month to get her out of the way.

Masts, funnels and various cabins were cut off the upright deck to clear the vessel of all her top hamper. Then the salvors, toiling night and day, built enormously strong tripods out of huge baulks of timber on the quay. By the time these were finished, lifting vessels were brought on the spot and moored close to the overturned ship. Cables were taken from the lifting vessels down under the keel of the ship and attached to the visible upper side of the hull, so the lifting craft, in straining upward, would tend to pull her over. Other cables were made fast to the deck and carried across the tops of the tripods on the quay.

Then came the touch of genius on the part of the Director of Salvage which makes the case unique. Five powerful railway locomotives steamed on to the quay and came to a stop by the sunken ship. The ends of the cables were made fast to the locomotives, and there followed one of the strangest tugs of war in the world between railway engines and a sunken ship. The five railway engines began to pull, and they pulled and hauled and strained away until they dragged the *Onward* upright. Pumping out soon followed, and within a month the scuttled troopship was raised and in dry dock. It was a difficult and novel feat, admirably performed.

It was by no means the first overturned ship that Commodore Sir Frederick Young had dealt with, for some years ago he righted and raised H.M.S. *Gladiator* after the *St.*

*Paul*, of the American Line, had crashed into her during a blinding snowstorm on April 25, 1908, and sunk her in the Solent. The British Admiralty called in the assistance of the Liverpool Salvage Association, who sent Captain F. W. Young, as he was in those days, to deal with the case.

Five railway engines hauling the overturned troopship upright. This extraordinary tug of war between a wreck and railway locomotives is unique in the history of the world

Up to that time it was as gigantic a task as any one had ever undertaken. There the cruiser lay on her side, 6000 tons of dead weight, on the sandy bed of the Solent, a fifty-foot hole ripped in her hull, several of her boiler rooms exposed to the sea, her grey plates just showing above the water.

The salvage expert was not a bit dismayed. He began to lighten the ship in every possible way. Her guns were taken out and salved. Then uncouth divers got busy with pneumatic chisels and cut off the funnels and ventilators and other deck fittings. Every hole in the deck was covered with wood and made watertight. Only the gash in her side, where the thick armour plates had folded down like tinfoil, was left open, and this in turn was dealt with by the divers, who carefully blasted away the ragged plates to prevent them from impeding the righting of the ship.

Pumping out the sunken troopship in order to raise her after she had been pulled upright by the railway engines.

Seven enormous pontoons, each 50 feet long, were made and lashed to the wreck. Two strong tripods were built up from the side of the hull, so that cables attached to the ends of the masts could be carried over them and hauled on by a couple of tugs when the time came to right the ship. The cables from the masts ran straight up in the air to the tops of the tripods, and when tugs began pulling, the tendency was to drag the ship over into an upright position. Inch by inch the *Gladiator* was turned after a terrific struggle, helped by 280 tons of iron which the salvors piled on the keel to press it down while the tugs were hauling up. The fight was severe, and even when she was righted her upper deck was still several feet under water, so the salvors determined to cover it with a huge coffer-dam built of strong planks. This coffer-dam looked like a great deck-house built up from the sides of the ship, and as it was made watertight and pumped out, it helped to pull the vessel to the surface.

Five months of strenuous work saw the pumps conquering the sea. The cruiser rose sluggishly, the tugs caught hold of her, and nightfall saw the little procession creeping into Portsmouth harbour. The cost of raising the wrecked cruiser was £50,500, and ultimately the Admiralty sold her to the shipbreakers for £15,125.

*By courtesy of the Merritt & Chapman Wrecking Company*
A very striking view of the overturned liner *St. Paul*, which provided some difficult problems for the American salvage experts.

The end of the *Gladiator* was the beginning of a dramatic sequel, a sequel so remarkable that it borders almost on the uncanny, raising once more the question whether there is anything in those legends of ghostly ships, like the *Flying Dutchman*, flitting about the seas until they are avenged or their long quest is over. For year after year the *St. Paul* sped along the sea lanes between America and England, thrusting through fog and shine and storm. Then the Great War demanded her conversion into a troopship, and early in the spring of 1918 the work was completed.

On April 25, 1918, ten years to the very day that she sank the *Gladiator*, the tugs were manœuvring her beside her quay in New York when she slowly began to heel over. Men gazed on her with amazement as she heeled more and more. Her masts touched the quay and crumpled like twigs, and as they smashed she went down on her side, even as the *Gladiator* had gone down in the Solent. In a short time 2000 tons of liquid mud gushed through her open portholes, which had now taken the place of her keel, and the salvage experts of the Merritt and Chapman Wrecking Company found her settled comfortably in a dozen feet of mud between the two quays. Why she sank is still a mystery.

*By courtesy of the Merritt & Chapman Wrecking Company*
Ten years to the very day after the liner *St. Paul* sank H.M.S. *Gladiator* in the Solent, she herself turned over and sank at her quay in New York. Sailors may be seen making a promenade of her hull the next day.

Mr. R. E. Chapman, the salvage engineer, had a most difficult problem to tackle. He had to grapple with a dead weight of 13,000 tons in a space so circumscribed that there was hardly room for the salvage craft to move. He did not worry. He set his squads of divers to work cutting away funnels and all the tackle from the top deck, as was done to the *Gladiator*, and when they had finished he sent them into the bowels of the ship in pairs in order to close all the open portholes that were buried many feet in the mud and over 50 feet below the surface of the harbour. It was inky black down below; they had no lights, because lights would not have penetrated the gloom, so they relied on their fingers instead of their eyes, and by using powerful hose to wash away the mud they managed to close over 500 openings in the ship.

One particularly clever piece of work was the making of a steel plate to fit over an opening around which were seventeen bolt holes. To get the bolt holes in the plate directly opposite the bolt holes in the ship seems almost an impossibility, but the diver solved the problem by taking down a sheet of lead which he hammered all round the opening until he had made a pattern with every bolt hole exactly in its place. From this pattern the steel plate was made, and it fitted perfectly!

Bulkheads to a ship afloat are an undisguised blessing, but the salvors found them a decided drawback on the sunken *St. Paul*. The bulkheads effectually stopped the flow of water from one end of the ship to the other, and before pumping could start it was imperative that the water should flow freely to the pumps throughout the whole length of the ship. It meant breaking through the bulkheads. The divers blasted through one or two with explosives, but the damage was such that the salvors decided to cut holes through the remainder with the electric torch.

Among the modern miracles that are little understood may be ranked that of creating a flame hot enough to melt metal immersed deep in the sea. Plunge a lighted match into water and the flame goes out; sink a blazing ship in the sea and the fire is conquered; yet the divers working on the *St. Paul* not only made a flame burn under the sea, but they also melted and cut holes through strong steel plates.

This marvel was worked by combining electricity and gas. The end of the torch was shaped like a cup, and the gas, driven at a high pressure through the pipe from the surface, reduced all the water within this cup to steam. Set in the centre of the cup was the electric terminal, and by holding it close to the metal plate to be cut an electric arc was formed with the terrific temperature of 6700 degrees! Under it the metal flowed like wax, and the divers were able to cut a dozen round drainage holes through the bulkheads. So blinding was the glare from the torch that even the muddy water was insufficient to stop it, and the divers were compelled to fit masks over their helmets in order to protect their eyes.

Meantime the men had been busy outside the ship, and there arose a long line of twenty-one legs, built of steel girders, all along the overturned hull. Shaped like the letter "A," 30 feet high, they presented a remarkable spectacle, and to gaze under their whole length was like staring at the under-framing of some mighty bridge.

Dredging a deep trench at the bottom of the next quay, the salvors sank twenty-one giant blocks of concrete, burying them with 15 feet of clay to make them immovable, and from these blocks they carried strong steel cables over the tops of the legs, and back to twenty-one steam winches set on the quay. When the time was ripe all the winches started to haul on the great legs, which began to lever the liner over. Powerful pontoons and wonderful floating derricks lent their aid, and after a ding-dong struggle lasting a week the liner came over sufficiently for the salvors to put in hand the final phase of the operations. Just as the *Gladiator* was floated at last by building a large coffer-dam over the deck, so the *St. Paul* was encased in a coffer-dam from end to end. Came a day when the pumps were set going, and the liner floated once more.

*By courtesy of the Merritt & Chapman Wrecking Company*
The wonderful maze of steel levers or legs, shaped like the letter "A," 30 feet high, erected on the overturned hull of the liner. by hauling on these legs with steel cables the salvors managed to drag the *St. Paul* upright

Salvage men are used to so much that they will tackle almost anything; but even salvage men would not tackle the 200 tons of decayed meat in one of the refrigerators of the liner. So horrible was the stench that they positively refused to go anywhere near. Money would not tempt them to the task. Eventually the trouble was overcome by a diver, who went into the refrigerating chamber fully equipped and was thus able to remove the carcasses without suffering from the offensive smell. It was a happy way out of the difficulty.

*By courtesy of the Merritt & Chapman Wrecking Company*
An excellent view of the *St. Paul* after she was raised, surrounded by the mammoth floating derricks which played so important a part in the salvage operations

While the experts will dwell upon the brilliant feat performed by the salvors in righting and raising the *St. Paul*, the average person will think of the strangeness of the case. That the liner should sink without cause on the tenth anniversary of the day that she sank the warship, that she should overturn like the warship, that pontoons, coffer-dams and legs erected on the hull should play so important a part in both cases, are all links in a chain of remarkable coincidences, the final link of which is provided by the fact that the salvage operations on liner and warship each took five months to complete. These are the incidents which make the case of the *St. Paul* so noteworthy.

The blizzard which caused the collision between the *St. Paul* and the *Gladiator* cost Great Britain a considerable sum, but not so much as the fog which led to the wreck of H.M.S. *Montagu* on the Shutter Rock at Lundy Island. The British Admiralty spared no effort or expense to get the battleship off, but after spending £85,000 in salvage work the navy had to confess itself beaten. So the proud battleship which cost over £1,000,000 was sold for the trifling sum of £4250 and was broken up for the sake of the metal she contained.

But for the genius of Commodore Young, the dreadnought *Britannia* might have met with a similar fate. Returning from a sweep of the North Sea during the war to her anchorage in the Firth of Forth, she was thrown by a heavy squall hard on the rocky island of Inchkeith. Tugs and torpedo boats failed to move her, and when Commodore Young came on the spot he found the rocks had not only pierced her bottom, but had also fractured her double bottom. Hopeless though her position seemed to others, the Director of Salvage considered it possible to refloat her.

All her stores, ammunition and coals were hauled out to lighten her. Still she sat tight, held firmly in the grip of the rocks. So a poultice of cement was fixed over the fractured plates in the second bottom to enable the engine-room to be pumped out, after which were made many connections leading into the flooded bottom. The air-pumps were linked up and set going, and as the air was driven into the flooded bottom it formed a belt which increased in depth until it expelled all the water through the holes made by the rocks.

Directly the salvors felt the battleship stir, they towed her off the rocks into dry dock, where the damage was quickly repaired. Duty called her later to the Mediterranean, where she was caught by a German torpedo and this time sent to the bottom for good.

# CHAPTER XIII

Of the many remarkable salvage feats performed during the war, that concerning the s.s. *Araby* is of more than passing interest. Driven ashore on the French coast on December 21, 1916, owing to an accident to her steering gear, she was towed off two days later and by Christmas Eve arrived at Boulogne. The tugs were shepherding the cripple into harbour when trouble overtook her once more. The towing hawsers parted, and she was swept by the strong tide broadside across the harbour mouth, her bow being jammed against the end of one quay and her stern against the end of the other quay.

The excitement was intense, for she was blocking our most important port of entry into France. To make matters worse, the tide was almost at the full, and unless she were got off at once it was obvious that her days were numbered. As the tide fell she was sure to ground at the bow and stern, and a deep channel between the quays left nothing to support her amidships, so she would be lucky not to break her back.

How the *Araby* blocked the entrance to Boulogne Harbour

As the tide fell, the *Araby* broke her back. This photograph clearly shows the fracture between the bow and the stern which led to her falling completely in halves.

Despite the utmost efforts, the *Araby* remained wedged between the two quays, and as the tide ebbed, her huge cargo of oats began to make its weight felt. Slowly she sagged in the middle until her keel was unable longer to support the strain. She broke her back and settled down right across the fairway, doing very effectively to Boulogne what the British Navy so gloriously succeeded in doing to Ostend and Zeebrugge.

It was a desperate case, calling for prompt measures, for somehow, anyhow, Boulogne harbour had to be cleared, and that quickly. Its urgency led to the happy co-operation of army and navy, so while the divers were jettisoning the cargo, in order to lighten the ship, Lieutenant-Colonel R. V. Jellicoe, D.S.O., of the Royal Engineers, was planning to make history by salving the first ship with the aid of ferro-concrete. Never before had anything like this been suggested. It seemed an impossible sort of dream.

The engineer was determined to prove that the seemingly impossible was possible. So on each side of the fracture, which was amidships, wooden moulds were deftly built up in the form of bulkheads stretching right across the inside of the ship. Cement and gravel were carefully mixed in certain proportions laid down by the engineer, and into these moulds the concrete was thrown. It set as hard as rock, forming two watertight walls shutting off the bow and stern of the ship, and leaving the fracture between them open to the sea.

The rapidity with which the work was carried out was so remarkable that by January 11, just eighteen days after the *Araby* was wrecked, the flooded compartments were being pumped out. To the joy of the salvors the rising tide lifted the ship clear of the bottom, and clever manœuvring enabled Captain H. Pomeroy, the salvage officer, to clear the harbour entrance and haul the ship into position practically parallel with the quay. By the end of the day she had been worked some little distance up the harbour and ships could pass in and out. The falling tide let her down again in the middle of the channel, but although she still interfered with traffic the salvors had carried the work a big step forward.

The hauling and the towing, however, had subjected her to a tremendous strain, as a result of which the crack across her keel began to extend up each side of her hull. This necessitated two strenuous days being spent in strengthening her, before she could again be pumped out and lifted a little farther into the harbour. Again she grounded at the fall of the tide, and once more as the tide rose she was lifted higher up the harbour. Throughout it was only possible to keep her afloat by continuous pumping, and once the pumps stopped she soon sank under the inrush of water.

Both halves of the *Araby* beached in Boulogne Harbour, where they lay for many months.

During these operations the crack had been creeping higher and higher up the hull under the alternating strains to which she was subjected. The mighty steel plates were rent and wrenched open until the greatest calamity of all overtook her and she broke right in two. She just fell apart, as a sliced apple falls apart, and sank to the bottom.

Towing the stern of the *Araby* back to England. the sight of half a ship afloat at sea is seldom seen.

Such a disaster would daunt most men, who would probably decide that the only thing to be done in so parlous a case was to finish the job by blowing the ends to smithereens and then to dredge up the pieces and throw them on the scrap heap. But the men tackling the case were in no wise disconcerted. If the problem had been complicated in one way, it had been simplified in another. For one thing, a ship breaking in halves required more delicate handling than one broken in halves, because the salvors would naturally try to prevent the worst from happening. Once the worst had happened, the salvors could go ahead without any thoughts of impending disaster. So, wasting no time, Captain Pomeroy brought some giant pontoons into play. Each was capable of lifting a weight of 800 tons, and by their aid, after a tremendous tussle, the two ends were lifted and beached out of the way of traffic in the inner harbour.

For weeks the tide washed in and out of them, leaving behind a foul sediment, and the remains of the *Araby* gradually became part of the landscape of Boulogne harbour—two ends of a broken ship, rusted and scarred, with the boilers in the engine-room exposed to sea and air. A year passed, during which the German submarine campaign kept the Salvage Section busy day and night, then the *Araby* was found to be interfering once more with our war activities. It was essential to extend the landing-place for flying boats and seaplanes at Boulogne, and the only available space was the strip of beach occupied by the two ends of the *Araby*.

In July, 1918, the frequenters of the harbour saw figures again at work on the wreck. The job of preparing the two ends to enable them to put to sea was carried forward with vigour. Then, unwittingly, came one of those tragedies which are fortunately rare in the annals of salvage. The ends still contained quantities of oats quite spoiled by the action of the sea. Grain in these conditions gives off fumes so poisonous that any one caught in them is instantly gassed and killed. Generally the fumes are kept down by spraying with chemicals, a procedure adopted during these operations.

One of the divers, however, penetrated too deeply into the hold without his diving dress and somehow got into a foul pocket of this gas. Almost at once he was overcome and fell in a state of collapse. No sooner had he fallen than his mate was also stricken by the fumes and rolled over unconscious.

Followed one of the gallant deeds which add fame to Britain's name. Discovering that the two men were in difficulties, and knowing full well the deadly danger that lurked below, a salvor lowered himself in an attempt to rescue them. Instantly the gas attacked him, and he, too, went down. By the time the three men were hauled out they were all dead.

Marred as it was by this sad tragedy, the work aboard the *Araby* was pushed ahead with unabated zeal. The concrete bulkheads, erected as described under the direction of Lieutenant-Colonel Jellicoe some fifteen months earlier, remained solid walls, impervious to the encroachments of the sea. So the Admiralty salvage officer completed arrangements for removing the remains of the *Araby*, and about the middle of July powerful tugs were hauling on the after end of the ship. At high tide they succeeded in towing the end off the beach into deep water, and the sailors of the

Dover patrol later witnessed the strange sight of half a ship floating serenely to England. They were more astonished a few days later to see the other half being towed across.

This torpedoed ship was the first in the world to be patched with concrete. The timber framework covering the hole in the hull forms the mould into which the concrete was poured.

The concrete patch from the inside of the ship, showing how the concrete was reinforced with steel rods.

In this wonderful way did a soldier, forsaking his own element, assist to salve a ship that broke in two, and so brilliantly successful was his work that he was "lent" to the Admiralty Salvage Section. On another occasion his genius was exercised upon a steamer which had a vast hole blown in her hull by a torpedo. Taking the case in hand, the soldier salvage officer determined to prove that ferro-concrete used under expert supervision would unite perfectly with the steel hull and make the ship as tight and sound as she had ever been. That concrete ships were possible was already proved, for there were one or two afloat to confound the sceptic, but the patching of a steel ship with concrete was not generally considered feasible.

However, the engineer set to work, and under his supervision divers built a huge mould over the gaping wound. The engineer himself donned a diving dress and went to the bottom to inspect the work and see that everything had been carried out to make the experiment successful. The concrete, reinforced with steel rods, was rammed into the mould, where it set almost as hard as the iron with which its edges were solidly united. Concrete piers were moulded inside the ship to strengthen the back of the patch and enable it to sustain the force of the waves, and when the vessel was pumped out and floated officials of the seamen's union, calling to inspect it, expressed their approval by certifying the ship as fit to go anywhere. It was an amazing new departure in salvage that proved an unqualified success. It was probably the first ship to be patched with concrete, although it was rumoured that the German cruiser *Goeben*, which gave us so much trouble in the Mediterranean, was also patched up with that material.

The *Araby*, however, was by no means the first ship to be salved in halves, for years ago Mr. Tom Armit, one of the cleverest salvage experts who ever tackled a wreck, undertook to recover the S.S. *Montgomery* which had sunk and broken in two in the river Garonne. Under his instructions divers timbered in the open ends of the vessel to make them watertight, and eventually each end was pumped out and raised. They were afterwards taken to dock and joined together again without the ship being one whit the worse for her adventure.

Equally remarkable was the salvage of the steamer *Milwaukee* which, going ashore on the rocks near Aberdeen during her maiden voyage in 1898, was held so securely that there was no hope of ever towing her off again. The salvors who were called in to deal with the case recognized this in a flash, but, gifted with a vivid imagination, they determined on an extraordinary experiment. It was the bow of the ship that was caught by the rocks, but all the valuable machinery was in the afterpart. Unable to save the ship whole, they made up their minds to try to save the half that mattered, planning to operate on the vessel just as a surgeon operates on a man, but, instead of using scalpels, they sought to cut with dynamite. A belt of dynamite cartridges was fastened round the ship just forward of the engine-room bulkhead. The brainy salvage men pressed the button. Scarcely had the sound of the explosion reached their ears when they saw the ship break in two and the stern slide into the sea.

How the concrete patch was strengthened with concrete piers on the inside of the ship to withstand the hammering of the sea

They had reason to be proud of their success, for it requires courage as well as imagination to operate on a ship in this manner. Eventually they towed the stern of the *Milwaukee* back to the Tyne, and in due course another bow was built and spliced on to the stern, thus making a new ship of her.

This noteworthy instance of ship surgery was duplicated in the case of the Atlantic liner *Seuvic* which went ashore on the Stag Rocks on the ragged Cornish coast. The untiring efforts of the salvors failed to move her, so they calmly cut her in two with dynamite and brought the after end to port, where she was made whole again!

a

View of the concrete patch in the ship's side after she had been pumped out

Those who get a living by marine salvage need be resourceful, masters of a hundred tricks to win ships from the grip of the sea. When the liner *City of Paris* came to grief on the same cruel coast, the jagged rocks cut right up through her hull and held her so tightly that her position from the first appeared hopeless. It seemed that she was destined to remain there hard and fast until the sea had battered her to pieces.

Whatever the underwriters thought, there was one enterprising salvage man who was prepared to match his skill against the strength of the sea. Offering to salve the ship on the "no cure, no pay" principle, he set his divers to work and little by little they blew away the rocks that transfixed the ship. It was a ticklish operation. Too strong a charge of dynamite would have injured the hull and made the case worse than ever; too weak a charge would have failed to remove the rock, so it was necessary to wed judgment with caution in this work. Bit by bit the rocks were blasted away and in the end the *City of Paris* was patched and floated. She was taken into Falmouth harbour for repairs, and when she again took the seas she was known as the *Philadelphia.*

That feat, performed a good many years ago, was equalled by Commander Cunningham of the Salvage and Towage Company when the Furness Withy steamer *Norton* ran ashore on Zogria Island off the coast of Greece a year or two ago. The rocks threatened to tear the whole bottom out of the ship if an attempt were made to tow her off, so the salvage expert, seeing there was no other way back to the sea, decided to blow the age-old rocks from beneath the bilges of the steamer. He set to work, and, using extraordinary judgment in placing the dynamite and gauging the power of the charges, succeeded in eight strenuous days in pulverizing the imprisoning rocks without doing any further injury to the steamer. At the top of the tide the tugs and salvage craft towed her into deep water and finally took her to port.

She was a rich prize, worth with her cargo some £330,000. The repairs to the steamer cost about £20,000, and the salvors by their fine work earned an award of £22,000. This seems a large sum for the salvors to make in so short a time, but it must be borne in mind that such prizes do not often come along, and the upkeep of a salvage steamer and her trained crew may easily run to £150 or more a week, without reckoning the cost of the steamer and plant, so it is plain that a big capital is required to keep a salvage unit in continual commission. In other words, although the award was good, taken in conjunction with the capital employed and the risk run, it was not by any means excessive.

# CHAPTER XIV

A ship cast ashore always reminds me of a hospital ward and the men and women who are deprived by illness of the power to carry on the struggle of life. The ship, too, is a cripple, driven out of her element, unable to carry on the duties for which she was created, and this is why my curiosity in a case is always tinged with a little sadness. To the salvage expert, however, the beached ship is merely a problem, and his mind, like that of the physician, is wholly occupied in effecting a cure.

If straightforward towing will not get the ship off, he will try other means. He may set a gang of men digging a deep trench round the keel of the vessel at low tide, and as the tide rises the water, flowing into this trench, will give her just enough buoyancy under her keel to enable the tugs to do the rest. Or he may try a trick that was tried very effectively on one occasion during the war when a whole convoy of ships grounded during a fog. The salvage officer, when his tugs failed to shift them, set torpedo boats thrashing round at a high speed and the wash they created lifted the grounded ships sufficiently for the tugs to get them off. It was a simple, yet clever, solution to the problem.

But there may be factors in the case which make these methods useless, as happened when the s.s. *Timbo* was thrown ashore in Carnarvon Bay in 1921. She drifted at the mercy of a terrific gale, which was blowing dead on the shore. Lifeboats that put out to succour her were swamped by the enormous seas, and more than one brave man lost his life that stormy day before the *Timbo*, absolutely helpless, was driven right across the bay. Just when tide and tempest were at their height, she was caught up by a tremendous wave and thrown heavily ashore.

That tide happened to be exceptionally high, and when Mr. Henry Ensor came on the scene he found a strip of shingle just 100 feet wide separated her from the sea when the tide was at the full. There she lay, broadside on to the ocean, and over 30 yards beyond the reach of the largest comber that rolled up the beach. She was indeed out of her element, so much so that 30 yards or 30 miles would have made no difference to the average city-dweller, for to him the problem of getting her back would have been insuperable.

To tow her off on a beach like that was not to be thought of, for if tugs had been set to work they would merely have added to the difficulties. Directly they began to haul, the stony beach would have heaped up under the weight of the steamer, and the more they pulled, the deeper the wreck would have burrowed into the beach.

The first thing the salvage expert did was to put timbers under the bilges of the steamer to prop her upright and prevent her from falling on her side. Then, using lifting jacks, he gradually raised her and placed launchways beneath her keel to prevent her from burrowing into the shingle when the tugs started to pull her off. This work was completed just before the highest tide there was likely to be for some time, and rather than miss this tide the salvors started to get the steamer back into the sea in the dark.

Inch by inch they hauled that steamer across the intervening shingle until half the space was covered, until the seas lapped the launchways, splashed the keel. It was a tremendous fight. The tugs were hauling to their last pound. Slowly the launchways disappeared into the water and at last the salvors felt the *Timbo* tremble. Another long, strong pull and the steamer rose to the swell. Success had crowned the efforts of the salvage specialist.

Refloating the *Timbo* was a fine piece of work, just as was the raising of the steamship *Fleswick* with compressed air by the same expert, many years ago. But in raising the *Silurus*, Mr. Ensor accomplished a feat that ranks with the finest wreck-raising feats ever accomplished. The *Silurus* was a dredger, one of the most powerful ever constructed. Built for duty in the port of Bombay, she was completed about eighteen months after the outbreak of war. As it was considered far too risky to attempt to tow her out to India at that time, she was taken to the Gareloch, where enemy submarines were not likely to penetrate, and anchored until such days as peace returned.

By digging a deep trench round this wreck, the salvors managed to tow her off into deep water

*By courtesy of H. Ensor & Sons*
The timbo, cast ashore a hundred feet above high water mark, where she was thrown during a terrific gale. Salvors propping up her bilges to prevent her from falling over before they started their stern struggle in the dark.

She had been serenely sheltered in that haven on the Scottish coast for nearly a year, when dirty weather sprang up. In the ensuing gale, she dragged her anchors and was driven hard ashore. Had she remained upright, a tug might have remedied the matter in a simple fashion when the tide rose again. But unluckily she grounded on a very steep shore, which shelved away rapidly, and as the tide dropped she capsized and buried her funnel so deeply in the mud that she was all but upside-down. The top of the tower carrying the dredging buckets was thrust into the bottom of the Gareloch, and while the tower tended to pull her over, once she had overturned, it no doubt prevented her from finishing with her keel right in the air.

As in the cases of the *Onward* and the liner *St. Paul*, the problem was to right the ship before she could be pumped out and raised. But with the *Silurus*, the difficulties were increased by the top hamper, consisting of the tower with the dredging buckets.

Mr. Ensor, as unlike a miracle-worker as any one could imagine, went to the Gareloch and quietly looked over the sunken dredger. She was a big problem, but not too big for him to tackle. Moreover, he had the courage to back his ability with his own money. Calmly he offered to salve the vessel on the usual "no cure, no pay" principle. It meant risking quite a fortune, but this did not worry him.

*By courtesy of H. Ensor & Sons*
The capsized dredger *Silurus*, with timber framing erected on her hull to prevent the steel ropes from cutting right through her

*By courtesy of H. Ensor & Sons*
The wonderful tangle of wire ropes and great blocks that were used to pull the *Silurus* on to an even keel again

Then he began to get out his plan for righting the vessel, the intricate calculations such a plan involves being not only amazing, but perfectly incomprehensible to the average man who is not possessed of engineering ability. He calculated on obtaining 1000 tons of lift by pumping compressed air into some of the compartments of the overturned vessel, and looked to pontoons attached to the tower and other parts of the structure to aid him in his plans. But, for the real work of pulling the ship over, he determined to rely on the power of steam-engines operating on the shore and hauling on a series of giant steel cables attached all along the ship.

The risk of pulling the ship to pieces in a job like this is so great that the novice would drag the ship apart far quicker and easier than he would drag it upright. If a cable were placed round the hull and a powerful steam-engine given full play ashore, that cable would crumple up the steel plates and gradually cut through them like a wire through a cheese, instead of moving the ship. These were the risks that had to be avoided.

Divers started to strengthen the ship with gigantic logs, 12 and 14 inches square, in order to withstand the terrific strain. A huge, strong frame of similar logs, protected by steel grooves, was fixed to the hull, to prevent the cables from cutting the ship to pieces.

It was slow work, for the salvors could only devote time to the wreck when there were no important war jobs to claim their attention. However, they managed to get in a day now and again, preparing for the great tug-of-war, upon which depended a fortune. Materials were not easy to obtain owing to the demand for munitions at the Front, so the salvors had to make shift with anything that would serve their purpose.

The divers, who set to work with hacksaws to cut holes through the steel plates for the passage of some of the cables, were greatly handicapped by the rust and mud, which made the water so cloudy that the work was difficult to see. Yet they stuck to their job and slowly, monotonously ate a way with their saws through the metal. Then they took up the task of preparing the seabed for the ship to come over on. She was practically lying on a submerged hill, and about a thousand yards of the seabed had to be removed to make a flat table on which the ship could rest in safety without slipping over again. All this took time as well as money.

Then it was necessary to find something ashore that

would withstand the pull of the ship when the tug-of-war started, something that would be absolutely immovable while nearly 2000 tons was dragging on the ends of the hawsers. The salvage expert tackled this difficulty by getting four old boilers, sinking them into pits dug down to the rock, and filling them and the space about them with concrete, thus making them as solid as the rock on which they stood. These boilers were in this way turned into four bollards, each capable of resisting a pull of 200 tons. Then a propeller shaft, 12 inches in diameter, was cut into suitable lengths and from it eighteen more bollards were made and set hard in concrete, each bollard being capable of withstanding a pull of 100 tons. These were placed at various intervals on the shore opposite the wreck, and by the time they were ready the salvors began to juggle with some 10 miles of steel cable, from 6 inches up to 8½ inches in circumference, that had been specially made by Bullivant, whose cables have dragged many a ship back into her element while making a snug sum for the salvors.

If there is any special work to be done, any heavy weight to be lifted, the salvage expert the world over knows he is safe with Bullivant's cable, that it will not break at the psychological moment and let him down. Some of these cables made of twisted strands of steel wire are 12 inches round—as thick as a man's leg at the calf—and they will support without breaking a weight of 320 tons: 320 tons could dangle from this cable in the air and a man could stand under it in perfect safety.

The largest hempen ropes made for salvage work are up to 24 inches round, even 25 inches on occasion, so it can be imagined how difficult they are to handle. If 1 foot of a 25-inch rope were cut off, it would be more than most men could lift, for it would weigh 146 lb. A short length of 15 feet would weigh practically a ton. A rope of this size will withstand a pull of 125 tons, against the 320 tons of a 12-inch steel rope. It might be thought that a rope half the size would support half the weight, but a peculiarity about hempen ropes is that, while a rope of 4 inches will support 4 tons, if you treble the size of the rope to 12 inches you increase the breaking strain by more than sevenfold to 29 tons; double the size of the rope again to 24 inches and it will support just four times the weight of the 12-inch rope, or 115 tons. Similarly, the bigger the wire rope, the bigger the load it will take in proportion. Whereas a 4-inch steel cable will support 35 tons, an 8-inch cable will carry 150 tons, or nearly five times as much, while a 12-inch cable will support 320 tons, or nearly four times as much as the 6-inch cable, which takes 88 tons.

*By courtesy of H. Ensor & Sons*
The *Silurus* raised, with the pontoons, which proved of the utmost assistance, floating near by.

Few people know that such wonderful ropes exist, but the salvage expert has full knowledge of where to get them when he requires them, as he did in the case of the *Silurus*. The ropes were all fixed in place on the edge of the Gareloch, two batteries of boilers were set up to supply the power, but before they could be used it was necessary to arrange a series of signals owing to the fact that the boilers were out of sight of each other. For one lot to haul faster than the other would have been fatal. It was absolutely essential that each rope took its share of the load and that all were hauled on at the same time. As showing how carefully everything must be considered in so important a case, the salvors even worked out how much efficiency they would lose through friction when hauling on the ropes. They left nothing at all to chance.

Giant wire ropes were lashed round some of the top gear to prevent it breaking away when the ship came over, a big trench was cut for one set of ropes to work in, as only by cutting the trench was it possible here to get a direct pull on the ship, and at last the signal was given to haul away.

Slowly the *Silurus* came up, her funnel was tugged from beneath 10 feet of mud. The hauling went on until the pontoons were clear of the water, until they were no longer a help but a hindrance, so the salvors cut through the wire lashings with blowpipes and freed them from the ship. Adjustments were made and the next haul set the *Silurus* on a fairly even keel. Despite the strain to which she had been subjected, the salvor made all his calculations so carefully that she was not in the least damaged by the operations. Over £56,000 was spent by the salvor on these operations, but he won his tug-of-war with flying colours, and the award he received was the reward of sheer merit.

As already mentioned, the divers used hacksaws to cut holes in the hull under water. In other cases they may bring into play a range of pneumatic tools—hammers, chisels, and drills worked by compressed air, which is pumped through a pipe from a boat on the surface. The hammer and chisel will deliver hundreds of blows a minute, each blow doing an almost imperceptible amount of work, but the hundreds of blows tell in the end. An air-driven drill, in spite of the disadvantages of working under water, will cut a hole an inch in diameter through a plate or girder an inch thick in one minute.

Frequently, it is desired to remove some submerged rock which interferes with navigation, and for this purpose pneumatic drills are often brought into play to make the holes for the charges of dynamite. The diver proceeds by drilling a series of holes, inserting his cartridges, after which he stops up the top of the hole with a special stopping in order to drive the force of the explosion downward. Then he withdraws to the surface and the boat removes to a distance before the dynamite is exploded.

Sometimes, however, when it is desired to deepen a rocky channel, a powerful rock-cutter weighing several tons is brought into play. This tool is shaped like a pencil and the nose is fitted with a specially hardened cutter. It is raised to a height and allowed to drop upon the rock, which it gradually pulverizes and breaks up, the rock-dredger coming along and completing the work.

Another method followed in the deepening of the channel of the Clyde was to use diamond drills for boring the holes for the explosives. The famous Enderslie Rock which caused all the trouble was revealed one day about the middle of the nineteenth century through the keel of a steamer coming into contact with it. Up till that time nobody knew of its existence, but when this steamer damaged herself the authorities started investigations. They found a bed of rock just over 900 feet long by 320 feet wide, which menaced the bigger ships that were beginning to navigate the river. The only way of making shipping safe was to deepen the channel by removing the rock. Accordingly it was attacked by men working in a diving bell who began blasting it away with gunpowder. By 1869, after working on it for five years and spending £16,000, half the channel was deepened to 14 feet, the other half remaining at 8 feet.

Eleven years later the rock was again attacked, this time by diamond drills worked by steam-engines. Five years of continuous work saw the rock removed to a depth of 20 feet over the whole channel. This improvement, which entailed the blasting away of over 100,000 tons of rock, cost £70,000, so the Enderslie Rock, upon which the Clyde authorities spent in all a sum of £86,000, proved rather an expensive obstruction to find in the river. But it was no mean feat to remove it, as was done, without in any way interfering with the traffic.

# CHAPTER XV

There have been few pluckier fights for a ship than that waged over a great, camouflaged merchantman torpedoed by the Germans off the Cornish coast during the war. She was badly holed, but her captain bravely stuck to her and managed to beach her near Bude.

Hastening to her aid, the salvage officer found her on a beach exposed to the full force of the Atlantic. With wind and sea rapidly rising, it was obvious that nothing could prevent her from going to pieces. The rollers were battering her, shaking and straining her ominously, seeking to finish what the German torpedo had begun.

So desperate was her situation that her one chance lay in reaching a more sheltered spot. The salvage officer looked at the sky, saw the wind blowing the crests off the waves, then he got busy. Working at pressure, he and his men managed to set a few baulks of timber within the ship to strengthen the damaged hull, and as the tide rose his tugs and salvage vessel started to haul her off the beach. He knew she was in a sinking condition, that she might go down before he could get her to a place of safety, but against this risk was her certain loss if she remained where she was.

Then began his struggle to beat the coming gale. The steamer was quite unmanageable, so he set two tugs hauling away in front, while he hung on behind with the salvage vessel, making his ship play the part of a rudder to the damaged craft. Along the coast northward the little procession made its way. The pumps were working continuously, throwing out tons of water, but they could not conquer the inrush. The captain and crew were still aboard, fighting hard to keep down the water. But all their efforts were useless. Gradually the ship sank lower and lower in the seas, and by the time they had reached Hartland Point—one of the most dangerous spots on that exposed coast—her end seemed but a matter of minutes. Her decks were practically awash. Heavy seas rolled right over them, and it became imperative to take off the men aboard. A dozen attempts were made in those heaving seas before the crew were rescued, and as the last man left he cast off the towing hawsers.

Only the *Ranger*, that famous salvage ship, hung on, still straining at the stern of the sinking steamer. A man stood by to slip the cable as she foundered, and the rescued crew crowded round to see her go, all waiting tensely for the end.

The ship which was given up for lost, after her men had been rescued with difficulty. The tugs, to avoid being dragged down by the foundering vessel, cast off their hawsers, but the salvage steamer still hung on to the stern and 7 gallant men gambled with death in a last effort to salve her

For a few moments the salvage officer watched the torpedoed ship. A few miles along the coast was Clovelly and safety. He wondered if he could make it in spite of everything, if there was yet a chance of snatching a victory over wind and wave, not to mention the Germans. After a close scrutiny of the ship, he determined to try.

Turning to his men, he called for volunteers to help him make one last attempt. Half a dozen men stepped forward. All knew the odds were against them, that a watery grave probably awaited them. Yet none hesitated.

In the face of incredible difficulties the salvage men triumphantly beached the sinking steamer at Clovelly.

Watching their opportunity, they brought their boat alongside the sinking ship and scrambled aboard. Then they took up the fight again. By great good fortune she had a donkey-engine on her upper deck, and the salvors succeeded in starting it up and getting the pumps working again. That donkey-engine proved their salvation, just enabled her to keep afloat. But it was touch and go all the time.

These seven gallant men in the end brought the ship to Clovelly harbour and put her ashore on that stony beach right under the picturesque village. She was nicely sheltered, and the salvors were able to fit her with a standard patch before taking her to dry dock. Thus the salvors wrested a victory out of the very jaws of defeat.

Several successful dramatists have staged a thrilling fight between divers, many a novelist penned vivid descriptions of similar encounters to make the hearts of his readers beat a little faster. Yet such struggles between real divers in the depths of the sea are so rare that it is doubtful if more than one authentic case exists.

This historic fight between divers took place at the bottom of the Solent during the recovery of some of the relics from the *Royal George*. The two divers, Jones and Girvan, were keen men, proud of their skill as submarine workers, each a little jealous of the other. One day Jones came across a cannon buried in the sand and, being unable to deal with it, marked it for a future occasion. Divers as a rule are extremely chivalrous. They would scorn to take a mean advantage, and they would never think of breaking the rule that what one finds, the finder salves. Whether Girvan, coming on the cannon, thought it a new find that he was entitled to salve, or whether he deliberately made up his mind to try to salve the other diver's find, is not known. All we know is that Jones, who had been working some little distance away, came on Girvan trying to get out the cannon. Naturally, Jones was indignant, and indicated to Girvan by energetic dumb show that the latter had no right to deal with the piece.

Girvan was by no means inclined to relinquish the cannon, and further remonstrances were followed up by blows. The divers began a rough and tumble fight at the bottom of the sea, striking at each other savagely with their fists. They were by no means equally matched, for Jones was much the smaller man of the two. Realizing that the encounter might cost him his life, he took the first opportunity of trying to get to the surface. Reaching the shot-rope, he went up it about 5 or 6 feet, closely pursued by Girvan who, grabbing his legs, did his utmost to pull him down again. The divers fought desperately in their rage, Jones to get away from those clutching hands that gripped his legs, Girvan to drag him to the seabed again, and that dramatic fight reached its climax in the greatest disaster that can overtake a diver. The glass of Girvan's helmet was smashed by a blow, and as the water swept in it seemed that his end was nigh.

Luckily, however, the men on the surface, unable to explain the violent agitation of the lines and feeling that something serious must be wrong, dragged both men to the top. Girvan's smashed helmet told its own tale and set them working frantically to pull him round. He was at his last gasp. Another minute and they would have been too late. He was removed to hospital, where his splendid physique, coupled with excellent nursing, enabled him to pull round. Those two divers who fought that strange fight at the bottom of the Solent came to the conclusion that it did not pay for divers to disagree, so they ended their differences by becoming the staunchest of friends.

Other attendants in tropic waters, feeling a strange dragging at the lines, have also drawn the divers to the surface without loss of time, to find them in the clutches of the deadly octopus, whose horrible tentacles have been coiling round the divers, striving to draw them within reach of the deadly beak that would go through the rubber diving dress as though it were paper. There, on the deck of the diving vessel, they have had to fight desperately to free the divers from the grip of the loathsome creature, only succeeding in the end by chopping and hacking away the encircling tentacles. As recently as the spring of 1924, when I happened to be in the South of France, a diver at Marseilles had to be rescued from an octopus in this thrilling manner.

The octopus, or squid, is, indeed, the greatest danger that the diver has to face beneath the surface of the sea so far as the denizens of the deep are concerned. Those squids occasionally found round the British coast are too small to threaten the diver, but in warmer waters, where the squid attains a huge size, he will rapidly attack any unlucky diver who unconsciously ventures too near his deep-sea lair.

The habits of fish are rather quaint. Should they be near the surface when a shadow falls on the water, a flick of the tail sends them disappearing into the depths. But undersea they are as inquisitive as cows. When fish see a diver standing still on the bottom, they find something about him too fascinating to withstand. Perhaps it is his form, perhaps the long line of bubbles flowing continually from the exhaust valve of his helmet. Whatever it is, they are drawn to the strange creature, and their fishy mouths suck at arms and legs and body in an effort to find out whether the diver is good to eat. The least movement sends them speeding away. The bigger fish are just as inquisitive, and just as easily scared. The diver needs only to open his air valve to let a little air escape in order to frighten them out of their fishy wits. Even the shark, the so-called tiger of the seas, is not generally feared by divers, for he is as scared by a sudden escape of air from the valve as are the smaller fish.

Yet the shark is fearfully inquisitive, and will come back again and again to see what the strange figure is doing.

Sometimes, indeed, the same shark becomes such a confounded nuisance, and the diver wastes so much time in scaring him away, that he is forced to put an end to the intrusion by slaying the monster. One diver, who had been worried day after day by the same shark, was compelled to signal to the surface for a knife. He then calmly held out his hand as bait, just as you hold out a bone to a dog, and as the monster turned to snap the delicacy, he stabbed it to death. Slipping a noose round the body of the fish, he sent it to the surface so that it would not attract other unwelcome visitors—for the scent of death in the sea is carried far afield by the invisible currents and soon brings the sea creatures swarming round—and was then able to resume his work in peace.

As already mentioned, it is often difficult for divers to see owing to the sand and mud suspended in the water, especially near the mouths of big rivers. A few feet down, and the light is quite shut out by the clouds of mud and sand floating about. Sometimes the divers work up to their armpits in foul slime—I recollect some years ago when a racing yacht was recovered from underneath 20 feet of mud—at other times the mud is so deep and thick that they spread-eagle themselves on its surface and manage to work in this recumbent attitude.

But when the diver gets to a hard bottom he is not handicapped in this way, and in sunnier climes and seas he can easily see at a depth of 100 feet. The sea-growths around Great Britain are not to be compared in size and colouring with the lovely tropic growths of coral and fern-like weed found in the warmer waters. Out, for instance, in the Pacific the depths of some of the lagoons are just like Fairyland: filmy forests of ribbons and ferns, inhabited by fish of the most gorgeous and dazzling colours, butterflies of the deep. This submarine scenery, in its way, is as beautiful as anything to be found on earth.

More than one salvage man in the past has made a snug fortune salving ships on the distant coasts of South America and the Pacific, often in the most simple manner by patching and pumping. Until comparatively recently the salvage man, if he wanted to lift a vessel, generally bought up a couple of old hulks and used these for slinging the wreck inshore. By the time the wreck was beached, the hulks were about smashed to pieces.

The principle of lifting a ship by means of a coffer-dam has already been indicated. It was a principle of which Mr. Tom Armit was a brilliant exponent. He raised several ships this way, building timbers all round to extend the hull upward, and then timbering all this over, virtually adding another deck to the ship. This coffer-dam, covering the whole ship, was made watertight, and, as it was pumped out, the added buoyancy refloated the ship. If leaks happened to manifest in the coffer-dam during pumping operations, the salvors calmly fed spun oakum into the water which carried it into the leak and soon stopped it!

On occasions during a collision at sea, mattresses and clothes have been thrown into the water, which has carried them to the leak, where they have become wedged, enabling the sailors aboard ship to tackle the damage from the inside. Collision mats are specially made for such emergencies so that they may be lowered over the hole, the pressure of the water holding them tightly against the side of the ship and enabling the carpenter to get to work on the inside as the inrush of water is stopped. Another salvor's trick is to stretch a tarpaulin over the hole to hold back the water. It is but temporary, yet it enables him to gain time to get timbers in place inside so that the pumps can then deal with the water that finds its way in. There are also special patches that may be pushed through the hole in the hull from the inside of the ship and opened out like an umbrella, after which they are drawn tightly against the hull by screwing up from the inside.

Pontoons alone have raised more than one little wreck in the manner already described. Other small ships have been raised by filling their holds with air-tight bags which, upon being blown up, have striven to rise to the surface, carrying the wreck with them, much to the delight of the salvors.

Vickers, the great armament firm, have their own patent system of raising wrecks by means of canvas containers. An American concern has a submarine machine, something like an army tank in appearance, for drilling holes in the hull of a sunken ship. These holes are drilled in line and large hooks are inserted, to which are attached strong, air-tight containers, one to each hook. The intention is to drill holes along each side of the hull of a wreck, attach the air bags, blow them up and lift the craft.

Whether the plates composing the hull of a ship are strong enough to support the entire weight of a ship in this way, or whether they would collapse under the strain of raising the ship from the bottom remains to be seen. It must be borne in mind that the backbone of a ship is the keel, that the whole ship is built up from the keel, which is its strongest part, the foundation of the ship. The inventors of this new system propose to lift the dead weight of the ship from the seabed, but hitherto salvors who have accomplished these feats have always swept their cables under the keel of the vessel to avoid the risk of pulling her to pieces.

Before the War there existed at least one special lifting craft, consisting of two steamers linked together by strong girders. These twin craft were brought into position so that the wreck lay between them, cables were fixed under the wreck, and the lifting craft picked up the sunken ship as the tide rose, steamed away with it until it grounded again, when the operation would be repeated next tide.

The salvors have several ingenious ways of getting cables into position. Sometimes two tugs towing cables between them sweep them under the wreck. At other times the end is let down to a diver who digs or scrapes a hole under the keel and forces the cable through; another rope is then let down from above, the diver attaches it to the end of the cable, which is drawn to the surface and attached to the lifting craft. A quicker method of forcing a hole under the keel is to use a powerful pump which, directed by the diver, rapidly drives a way under the wreck for the lifting cable.

It was while using a pump for this purpose on the wreck of the *Intrepid* on the Belgian seaboard that a most amazing adventure befell a diver of the Salvage Section. The wreck was buried 20 feet in clay and mud, and the diver by skilful use of the pump dug his way down to the keel. He was standing at the bottom of this pit when it caved in on top of him. He was buried alive, held as in a vice under a dozen feet of mud and clay, the weight of which doubled him up.

Luckily he still retained his hold of the pump, and after a desperate struggle managed to direct the jet of water on to himself until he loosened one arm. As the water softened the clay, he worked the other arm free, then little by little his legs. Wrapping them round a wire, he directed the pump upwards and inch by inch wriggled and burrowed his way through that dozen feet of clay to the surface. His air-pipe was hopelessly entangled, so he was compelled to cut it before he could be hauled up to safety. No diver would care to undergo such an experience a second time.

Comedy so seldom plays a part in diving adventures that a case which occurred some years ago is worth recording. Divers had been at work for some time hauling the cargo out of a submerged wreck, when one of them, upon being drawn up, displayed quite exceptional signs of exhaustion. A sleep soon put him right, and he resumed work next day.

Again he showed signs of acute fatigue, which passed away after a night's rest. The following morning he went down as usual, and this time when he came up he was quite unable to stand. He collapsed on the deck, while those aboard crowded round, very concerned about his safety.

Hastily unscrewing his helmet, one of the salvors sniffed in a puzzled sort of way. A familiar smell came to his nostrils. He sniffed once more, the others looking at him queerly.

"What's wrong?"

"Whisky!" muttered the kneeling man, thinking his sense of smell must have betrayed him.

They all sniffed in unison, and the smell was unmistakable.

"He's drunk!" said the first man.

The idea was preposterous!

"But how——?" queried another.

That was the question which baffled them. How was it possible for a diver to get drunk under water? The mystery would have delighted Sherlock Holmes. There were cases of whisky in the wreck at the bottom of the sea, but the diver would be drowned if he attempted to drink it. He was imprisoned in his suit. So how?

Not a word did they say to the drowsy diver, but when he went down the following day another diver discreetly followed. He saw the first diver take a bottle of whisky and proceed to a cabin. Instantly the mystery was cleared up. The exhaust air from his helmet, collecting here, had formed an air pocket, and the diver, poking his helmet out of the water, calmly unscrewed the glass front and took a good pull at the bottle. In this ingenious manner did he manage to get drunk under water!

For recovering metal objects, such as anchors accidentally lost in dock, there is the electric magnet. Among other inventions for seeing on the seabed and recovering lost treasure is the hydroscope of the Italian, Cavaliere Pino. The hydroscope is a floating chamber, from which depends a series of steel pipes that may be extended or shortened at will, just like a telescope. The pipes terminate in a chamber with observation windows made of stout glass, and a man sitting here can observe the whole seabed round about, provided the water is clear, while the hydroscope is being slowly towed along on the surface.

The hydroscope has done some good work, and by its aid one wreck was raised in five hours after salvors who had been working on it for months had declared that the craft was lost for ever. It was this Italian invention that the Japanese used in clearing the sunken Russian fleet from the bottom of Port Arthur after the termination of the Russo-Japanese War in 1905. A similar invention worked out by a Mr. Williamson has resulted in some extraordinary underwater cinema films being secured.

The War led to a big development in the use of compressed air for raising wrecks, divers sealing up all the apertures in the tops of the wrecks with concrete to imprison the compressed air, which was then pumped into the ship until enough water was expelled to enable her to float. The War also hatched a crop of cranky salvage ideas that gave some of the salvage experts one or two happy moments.

One such moment was just after the War, when an American walked into one of the British shipping departments and requested to be allowed to salve a ship in order to demonstrate the efficiency of his new method. The officer to whom the stranger went was courteous, listening attentively to the American's demand, and inquiring at last which ship of the few hundreds sunk round our coasts he would like to demonstrate on.

When a ship overturns on quick sands, the salvors erect great legs on the hull, as shown here, and take strong steel cables from the masts of the wreck over the tops of these legs and haul on them until they drag the ship upright.

"Any one!" said the American. "I don't mind. The bigger the better. What about the *Lusitania*?"

"She's rather deep," it was suggested.

"That doesn't matter. It makes no difference to me what the depth is," came the easy reply.

The officer put a few questions, and then learned that the stranger designed to use a submarine, which was to fire torpedoes right through the *Lusitania*, each torpedo carrying with it a steel cable. These were to be picked up at the other side and taken to the surface, and then the wreck was to be dragged bodily out of the depths!

That scheme to salve a ship by first of all smashing a series of holes through her hull with torpedoes did not commend itself to the British expert. It was, indeed, quite impracticable.

None the less, there are people who still wonder if it will ever be possible to salve the *Lusitania*, which was torpedoed off the Irish coast on May 7, 1915. From time to time the matter keeps cropping up.

Those who are curious on the subject may be interested to know that the chances of raising the *Lusitania* are so small as to be almost negligible. The sheer weight of the sea quickly obliterates man's handiwork, and the *Lusitania* probably ceased to be a ship years ago. It is extremely likely that the tremendous pressure to which she was subjected at the depth of 288 feet long ago crushed her flat. Proposals have been made to try to salve the valuable 30-ton safe from the strong-room of the liner, but personally I should not care to back such an enterprise.

The marvellous endurance of divers in going to great depths has been touched on in previous chapters, but perhaps the strangest task ever given to a diver was that of saving a cathedral. Some years ago, Winchester Cathedral was in such grave danger of collapsing that it became necessary to underpin the walls and strengthen the foundations. The whole cathedral stood upon a water-logged peat bog, the ancient builders upon reaching water having laid logs of beech to take their foundations. The modern architect, Mr. T. G. Jackson, and his engineering collaborator, Mr. Francis Fox, knew that to pump the water out would be practically to pump the cathedral to destruction, for the drift of the water was bound to carry the silt and gravel away from other portions of the building to where the pumps were working, and so bring about the collapse of the famous edifice.

After careful study of the difficulties, the engineer called in one of the crack divers of Siebe, Gorman & Company to carry out his plan. It was found that the beech logs put in by the ancient builders at water-level were resting on 6 feet of clay, which in turn covered a depth of just over 8 feet of peat, this in turn resting on a bed of gravel. To save the cathedral it was essential to excavate all the clay and peat down to the gravel, and replace it with concrete up to the foundations of the building.

The walls of the cathedral, properly supported, were treated in small sections of about 5 feet. The clay was dug out, then the diver entered the hole and, working in absolute darkness, removed the peat down to the level of the gravel. Bags of dry concrete were lowered to him and packed in tightly, a layer at a time, the diver splitting them open and spreading the contents evenly. In this way the hole was completely filled. The water soon turned the concrete into a rock-like mass, upon which the masons were able to build solidly right up to the foundations, from which the beech trees were carefully removed. Nothing like it was ever attempted before, so Winchester can boast that its cathedral is the only one in the world that has been given a solid foundation by a diver.

Just as the torpedoing of the *Lusitania* by the Germans stirred the whole world, so the sinking of the American flagship *Maine* in Havana harbour on February 18, 1898, stirred the people of the United States and led to the war with Spain. A giant explosion in the middle of the night carried the American battleship to the bottom with 266 officers and men, and it was asserted that the Spaniards had deliberately blown her up. The result was a war in which Spain lost Cuba and the Philippines.

Long years afterwards, in 1910, Congress voted a sum of £60,000 and the work of investigating the wrecked battleship was put in hand. Tackling their task in a most masterly manner, the engineers decided to enclose the whole wreck in one huge coffer-dam built of steel piles driven down through the mud until they were embedded 13 feet in the solid clay. As the wreck lay in 37 feet of water, with 20 feet of mud below that, the piles would emerge 5 feet above the surface of the sea, providing a wall too high for the water to wash over.

Knowing full well that they would find it difficult to create a plain circle of piles round the ship to withstand the pressure of the sea, the engineers decided to build what really amounted to a series of gigantic barrels, standing on end in the sea with their sides touching. These barrels, twenty-two in number, varied between 40 feet and 50 feet across. The staves of the barrels were formed by the steel piles which were made to interlock as they were driven in side by side, and where the barrels, or caissions, touched, further piles were driven to enclose the space and strengthen the junction.

For months the hammer-blows of the pile drivers resounded over the harbour, and at last the coffer-dam—a most marvellous piece of work—was finished and filled with dredged clay. Within a year the salvage operations were completed at a cost of £135,000. The experts watched with keen eyes as the pumps lowered the water within the coffer-dam and the wreck slowly emerged from the slime. There the battleship lay, a twisted mass of metal, and, before patching up the afterpart and taking it out on March 16, 1912, to bury in the broad Atlantic, the specialists held their inquest, striving to discover whether the explosion that sank her was caused from inside or outside.

Such a thing after a ship has been at the bottom for over twelve years is almost impossible to determine. It was said that the explosion came from outside, but the doubt will always exist that the Spanish American War may have been due to a grave error on the part of America, and that the *Maine* instead of being blown up by the Spaniards, was destroyed by the spontaneous combustion of the explosives in her magazines, just as French, Japanese and British warships have been destroyed in the same accidental manner.

THE END

www.ingramcontent.com/pod-product-compliance
Lightning Source LLC
LaVergne TN
LVHW022220200926
848481LV00006B/2048
* 9 7 9 8 3 4 9 2 1 7 7 0 8 *